St. Louis Community College

Library

5801 Wilson Avenue
St. Louis, Missouri 63110

Satellite Broadcasting

Satellite Broadcasting

P. RAINGER
D. GREGORY
R. V. HARVEY
A. JENNINGS

A Wiley–Interscience Publication

JOHN WILEY & SONS

Chichester · New York · Brisbane · Toronto · Singapore

Library of Congress Cataloging in Publication Data:
Main entry under title:

Satellite broadcasting.
  'A Wiley–Interscience publication'.
  Includes index.
  1. Direct broadcast satellite television.
I. Rainger, P.
TK6677.S28   1985   621.388'5   85-9366
ISBN 0 471 90421 X

British Library Cataloguing in Publication Data:
Satellite broadcasting.
  1. Artificial satellites in telecommunication
  2. Broadcasting
I. Rainger, P.
621.38'0422   TK5104

  ISBN 0 471 90421 X

Printed in Great Britain

Preface

There are four authors to this book:

David N. Gregory, MA.,C.Eng.,FIEE.
Robert V. Harvey, BSc.,DFH.,MIEE.
Antony Jennings, MA.
Peter Rainger, CBE.,FRS.,F.Eng.,BSc(Eng).,FIEE.

I have edited the contributions of these authors to remove duplication or inconsistency in the areas where they overlap. I have also added material, where necessary, to remedy omissions so that it would be neither fair nor accurate to attribute sections of the book to individual authors.

The book covers a wide range of subjects related to Direct Broadcasting Satellites. It cannot pretend to be a specialist book on any of these subjects. Rather it is intended to give a specialist in one field a broad understanding of the contributions being made in other fields. Those who require further details should consult the documents listed in the references at the end of each chapter.

Alternatively the book may be read by the non-technical person who may wish to skip the more difficult parts. For these people a small amount of duplication was thought to be helpful to make each chapter independent.

The book depends on knowledge gathered from many sources and the authors are particularly indebted to colleagues within the British Broadcasting Corporation and British Aerospace.

On behalf of all concerned I would add that the views expressed are, nevertheless, those of the authors and do not represent the views of these organisations.

In addition much information originated from other broadcasters, notably the European Broadcasting Union, or is abstracted from documents of the International Telecommunication Union, and their contribution is gratefully acknowledged.

To many experts the chapters dealing with their specialism will fail to do justice to the subject. Simplification of some more complex subjects has been attempted to limit the size of the book but the authors hope that inaccuracy has been avoided and that the most important points remain.

It was the authors' intention to write a book which was international in character. However, information is not available to give a totally international balance to the subject. Thus the book tends to accent the European situation and, in particular, that of the U.K. However the fundamental issues have worldwide application.

To those who feel their national contribution has been neglected we offer an apology.

Peter Rainger
Editor

Contents

CONTENTS (continued)

Appendices

Introduction

Satellite broadcasting is simple in concept: it is the
practical implementation which presents the problems.
In concept the satellite is only a repeater which
    receives modulation on one frequency and re-transmits
this modulation at higher power on another frequency.
Satellite communications use a transmitter on the ground
which beams a signal up to the satellite. The satellite
receives this signal and re-transmits is back to earth
so that it may be received throughout the country.

It is well known that high frequency broadcast signals
travel in straight lines and the signals cannot be
received beyond the horizon. Thus a transmitter
theoretically serves a circular area around the mast
(although in practice it serves a more irregular area
because the earth is not a perfect sphere but has hills
and dales).

The size of this circle is determined by the height of
the mast. A broadcasting satellite could be considered
to be a transmitter on a very high mast. As a result
the transmitter can illuminate the valleys and there are
no unserved shadows and there is no limitation at the

horizon. If the transmitter has sufficient power it
could broadcast to the world (or rather the half that
faces the satellite). As we shall see later the broadcasting satellite is normally required to serve only a
small area and the transmitter beam is confined to a
small cone.

It follows that broadcasting by satellite is attractive
because the whole population is served immediately by
the one transmitter. This is, of course, a slight
exaggeration because there will be a small number of
people who cannot see the south west sky or cannot erect
an aerial in a suitable place.

Nevertheless there are many technical problems to be
overcome before such a broadcasting system can be
constructed.

This book is really the story of how these practical
problems have been overcome. How do you build a DBS
transponder into a satellite? How do you get the
satellite up there? How long will it last? What quality
of service will it provide? What domestic receivers are
necessary?

These are just some of the many problems to be solved
before you have a broadcasting system.

Even when all this technical work has been done it is
desirable that we use this valuable resource in an
orderly manner. We need a regulatory framework and we
need to ensure that we can operate within the legal
system of all countries. There are regulatory and legal
problems to be overcome, new regulations to be devised
and new laws to be enacted.

Last, but not least, we need to consider the financial issues. Can we pay for it? How long will it last? Is it a worthwhile investment?

This book is devoted to answering some of these diverse questions.

4

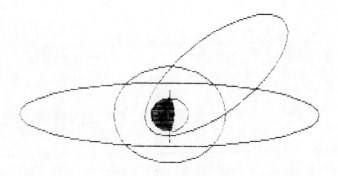

Fig. 1.1 Path of satellites orbiting the Earth

CHAPTER 1

Celestial Mechanics

1.1 HISTORICAL SURVEY

History contains many examples where a theoretical understanding was a necessary prelude to practical exploitation. In one sense celestial mechanics was an exception because, in the early 17th century, an astronomer named Kepler published his "Laws of Planetary Motion" and they were based on careful observation rather than a theoretical understanding of the subject.

He had discovered that the movement of the planets appeared to obey the following laws:

1. The planets have an elliptical orbit with the sun at one focus
2. A line between the sun and an orbiting planet sweeps the intermediate area at a constant rate
3. The square of the orbit period is proportional to the cube of the major axis of the ellipse

Newton (1642-1727) subsequently established his "Laws of Motion" which showed that a mass (m) has an acceleration (f) when subject to a force (P).

6

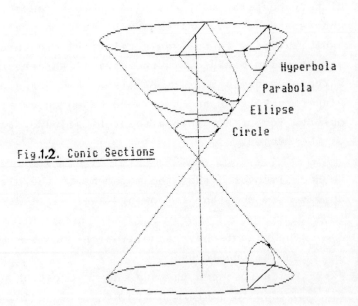

Hyperbola
Parabola
Ellipse
Circle

Fig.1.2. Conic Sections

After carefully observing the motion of the moon he proposed his "Universal Gravitational Law" which said that the gravitational attraction between two objects was proportional to the product of their masses and inversely proportional to the square of the distance d between them.

He then substantiated Kepler's Laws for captive bodies.

The motion of the planets was a favourite subject for study in the years that followed. Laplace (1749-1827) calculated the motion of the solar system and Gauss (1789-1857) put the finishing touches to the study of the orbits of celestial bodies.

In more recent times relativity studies have added to this picture of the heavens but this concept is not significant in the examples which we will study.

We are interested in the special case of a small mass moving about a large one (the Earth) with a known velocity. It can be shown that its trajectory is one of a class known as conic sections. These are the circle, ellipse,parabola and hyperbola. The first two represent captured bodies, that is bodies in orbit about the Earth and the remaining two will eventually escape from the Earth's influence.

Add the concept of constant total energy (kinetic and potential) and constant angular momentum and we have a system which obeys Kepler's Laws. Using the elementary equations of Newton we can calculate all we need to know for the purposes of this study of Direct Broadcasting Satellites.

While a man-made satellite would not have seemed
theoretically impossible to these early mathematicians
it would have seemed impractical because the energy
required to launch the satellite was beyond their
wildest dreams. The second World War with its ballistic
missiles was the motivation for a leap forward in the
development of rocket motors. Before long we had the
Intercontinental Ballistic Missile and we had motors
that were capable of launching a satellite.

In October 1957 the first satellite was launched
(Sputnik 1) by the USSR. Unfortunately it only lived
three weeks in a favourable low orbit but it was a
beginning. The first communication satellite (Score)
was launched by the USA in 1958 but this too had only a
short life.

Somewhat surprisingly it was not until 1946 that Arthur
C.Clarke drew attention to the possibility of a
geostationary satellite. This was a satellite in
circular orbit around the Earth with an angular velocity
that matched that of the Earth. Such a satellite would
appear to be stationary in the sky and this would
greatly simplify the design of transmitters and receivers
communicating with the satellite. Unfortunately such an
orbit required a very powerful rocket motor but progress
was such that, by 1968, Canada expressed anxiety about
the possible overcrowding of the geostationary orbit.
(It is thought that 2389 launches have been made during
the years 1957-1982). In response to this the World
Administrative Radio Conference was planned and the basis
laid for the DBS services of today.

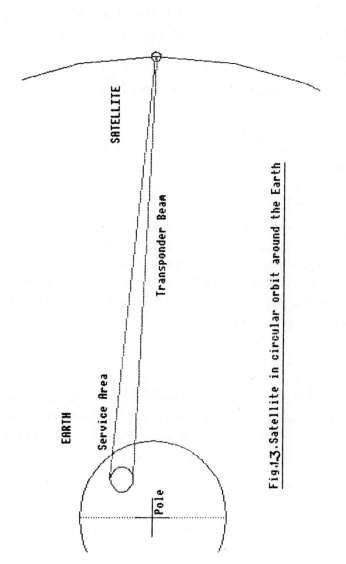

SATELLITE

EARTH

Service Area

Transponder Beam

Pole

Fig.1.3.Satellite in circular orbit around the Earth

1.2 <u>ELEMENTARY DYNAMICS</u>

First we need to establish an understanding of some
fundamental concepts. We all know that it is necessary
to use force to move a mass (or more accurately,
accelerate it) and the mathematical expression of
this principle is often described as one of the Laws of
Motion:

    Force   = mass   x acceleration

This acceleration gives the body a new velocity and
changes its momentum and kinetic energy.

When the point of application of this force moves, work
is done and the body acquires kinetic energy. We will
all be familiar with the concept that a body can have
both kinetic and potential energy. From our earliest
years we recognise that the potential energy of an
object is converted to kinetic energy as it falls to
earth and, to our dismay, we often find that this energy
is enough to break a fragile object as it hits the
ground.

There are other forms of potential energy. The rocket
designers are particularly interested in the energy
encapsulated in a chemical bond and, as we shall see
later, the process of launching a satellite involves
changing one form of energy into another.

In the isolation of deep space changes in total energy
can only occur because of radiation. Thus we have the
concept of "conservation of energy". Similarly an
isolated system must have "constant momentum".

The other important factor is the gravitational
attraction between two masses. We are interested in
the special case of a small object in the vicinity of
a large one and for simplicity we assume that this body
is spherical.

Let us consider first the circular orbit. We have a
special interest in the orbit whose period is one day
because, if it is in the equatorial plane and rotating
in the same direction as the earth, the satellite will
appear to be stationary in the sky.

It should be noted that velocity is a vector quantity,
that is, it has both magnitude (speed) and direction.
Newton's Law asserts that is necessary to apply a force
if the velocity of a mass is to be changed. Thus, even
if the speed is maintained constant, a force is
necessary to change the direction of motion.

If we spin a weight around us on a piece of string we
know that the weight will try to fly off the end unless
we restrain it. Although the weight is travelling at
constant speed the direction, and hence the velocity,
is constantly changing. The restraining force makes
the weight accelerate towards the centre so as to
maintain its circular path. If we remove this force
the weight will fly off into the distance, travelling
at constant velocity. If we pull harder on the string
the weight will move towards us.

The rotating mass is often considered as having two
forces in equilibrium. The centrifugal force acts upon
a body in curvilinear motion in a direction which appears
to be pushing it away from the centre. The centripetal

force is the restraining or balancing force that actually
causes the body to accelerate towards the centre and
hence rotate in a circle.

In the case of the satellite the piece of string
providing one of these forces is replaced by gravitational attraction and the satellite orbit is stable when
these two forces are equal.

If the satellite is in a circular orbit then it can be
shown that:

Centrifugal Acceleration = Angular Velocity

x Radius

We are all aware that every mass has weight. This
weight is the gravitational force that is attracting
the mass towards the earth and, if there is nothing
supporting it, it will accelerate towards the ground.
This acceleration is given the symbol 'g':

$$\text{Gravitational Attraction} = \text{Constant } (k^2) \text{ x } \frac{\text{Product of masses}}{(\text{Separation of masses})^2}$$

The centrifugal and gravitational forces act in opposite
directions and luckily for us the gravitational forces
are much the stronger. However if we consider the
situation at an altitude of many kilometres we find that
the relative magnitudes have changed. The centrifugal
forces have become stronger and the gravitational forces
are much weaker. At a critical height these two forces
will be equal in amplitude and the satellite will be
in equilibrium.

Calculating on the basis of unit mass we have for equilibrium between the centrifugal and gravitational forces:

$$w^2R = \frac{k^2m}{R^2} \qquad \dots\dots\dots\dots (1)$$

where: 'w' is the angular speed of rotation about the earth

'R' is the radius of the orbit

'm' is the mass of the earth

and $w = \frac{2\pi}{T}$

where: 'T' is the period of the orbit

Now $\frac{k^2m}{R_o^2} = g$

where: 'g' is the gravity acceleration at ground level (R_o)

Substituting in equation (1) we have:

$$w^2R = g\left(\frac{R_o}{R}\right)^2$$

<u>Kinetic energy</u> is $\frac{1}{2}V^2$ ($= \frac{1}{2}w^2R^2$)

where 'V' is the speed of the satellite

As the satellite rises the work done when moving against gravity is converted into potential energy. If the satellite (at radius R) is sent a very large distance away it will escape from the earth's influence and the increase in potential energy is as follows:

$$\underline{\text{Potential Energy}} = \frac{km^2}{R} \quad (= w^2 R^2)$$

for unit mass

Thus comparing the two equations we have shown that if the orbit is stable and there is a balance between the two opposing forces, then the <u>kinetic energy must be half the potential energy as defined here</u>.

Note that the potential energy of a mass, at a given altitude, is defined as the energy required to escape from the earth's gravitational field. (Conversely it is the energy that is released when a mass falls to a given altitude).

This definition of potential energy is adopted because it simplifies the algebra of the calculation and it has more direct application to objects in orbit around the earth.

Conventionally potential energy is defined with respect to the surface of the earth. The relationship between potential energy as defined in these two alternative ways is, of course, a simple one.

It follows that the total launch energy requirement is

$$k^2\, m\, .\left(\frac{1}{Ro} - \frac{1}{2R} \right)$$

This energy has to be provided by the launch vehicle
and it will be seen that the total energy required for
the low orbits used in the early days of satellites is
only about 1/2 that required for a geostationary orbit.
(Furthermore if the orbit is in the right direction
the rotation of the earth itself can give the satellite
a flying start).

Nowadays sufficient energy can be provided to realise
a geostationary orbit and, if required, a complete
escape from the earth's influence can be achieved for
only a small further increase in energy.

Finally each stable circular orbit has a unique period
of rotation and we can deduce from these equations
that:

$$R^3 = 10.104\ \times\ 10^{12}\ \times\ T^2$$

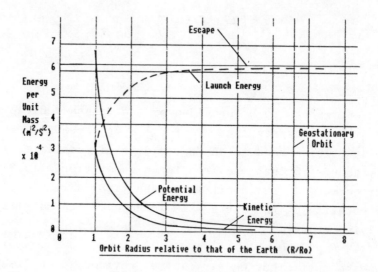

Fig.1.4 Orbit Energy (per kg)

The following constants were used in the calculation:

k^2 the Newtonian gravitational constant is 6.670×10^{-11} $(m^3 kg^{-1} S^{-2})$

Ro   is 6.378×10^6 (m)

m   the earth's mass 5.976×10^{24} (kg)

g   gravitational acceleration at the earth's surface 9.806×10^{-3} $(m S^{-2})$

This function is Keppler's third law (see Fig 1.5).

For the geostationary orbit T =   24   hours   and the radius of the geostationary orbit is approx 42,000 Km.

If we push a satellite up to this altitude and give it the required velocity it will rotate once every 24 hours. Furthermore as there is no air friction it should, according to this simple theory, stay there indefinitely.

As we shall see later other orbits are possible which have different properties.

1.3 <u>LAUNCH MECHANISM</u>

A key question when considering which orbit to employ is the total energy that it requires.

It should be noted that a small increase in the orbital height becomes very demanding in terms of the launch weight because of the fuel which must be carried in the   early stages.

This preoccupation with total energy led to the consideration of multiple stage launches. This, in turn, forced careful choice of the most efficient launch programme and Fig.1.6 is a theoretical solution known as the "Hohmann" type transfer. In this case the satellite is first launched into a low circular parking orbit. Subsequently it is put into an elliptical transfer orbit and, when it reaches its apogee, a third burn puts the satellite into the required high circular orbit. In practice a number of other factors will influence the choice of the most efficient launch programme.

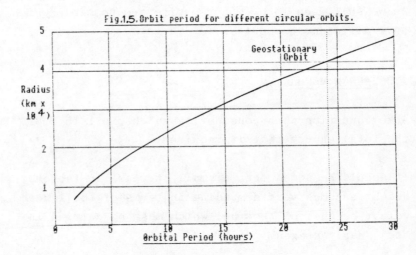

Fig.1.5. Orbit period for different circular orbits.

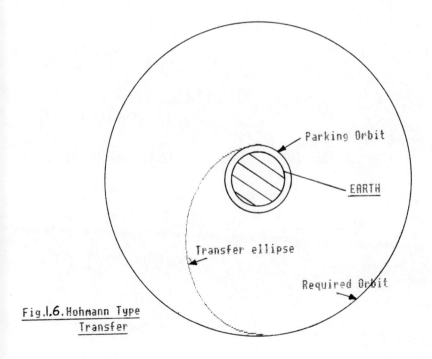

Parking Orbit

EARTH

Transfer ellipse

Required Orbit

Fig.I.6.Hohmann Type
Transfer

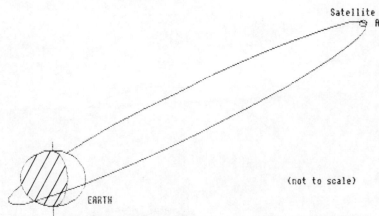

Fig 1.8. Satellite in elliptical orbit which is quasi-stationary
at the point A

1.4 ELLIPTICAL ORBITS

The elliptical orbit is an interesting orbit because it requires relatively little energy and, if launched in a non-equatorial plane in a non-synchronous orbit, it will scan the earth in a way which is very useful for survey purposes. However it requires a tracking antenna and so is not normally considered for satellite broadcasting.

However, Kepler's second observation suggests that the satellite will be travelling slowly when it reaches its apogee. If we have an elliptical orbit such that, at one extreme the satellite swoops low over the earth,then later,when it retreats to distant space,it will appear to be nearly stationary for many hours. If we choose the orbit carefully we can have a broadcast transmitter overhead for a significant portion of the day. It is then particularly easy to align the receiving antenna and it could be useful for mobile communication.(Fig.1.8)

This concept is employed by the "Molniya" series of satellites launched by the USSR. Unfortunately this type of orbit suffers from minor disturbances such as "apsidal rotation" and "nodal regression" which require the use of additional fuel to keep the satellite on station. (Apsidal rotation is the rotation of the major axis of the ellipse within the plane of the orbit. This is zero when the orbital plane is at 63 degrees. Nodal regression is the rotation of the orbital plane).

The apparent path of this type of satellite, as seen by an observer on earth depends on the orbital period which may be a multiple (or submultiple) of 24 hours. A typical path around the world where the satellite appears to pass directly overhead is shown in Fig.1.9.

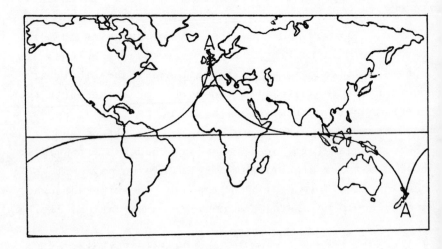

Fig.1,9 . Elliptical Orbit quasi - stationary

over the point A

- period 12 hours

CHAPTER 1

References

No.	Subject	Book Reference
1	The RAE Table of Earth Satellites 1957 - 1982	1.1
	D.G.King-Hele,J.A.Pilkington, D.M.C.Walker,H.Hiller,A.N. Winterbottom	
	Publisher - Macmillan, London	
2	Men of Mathematics	1.1
	E.T.Bell	
	Publishers - Simon & Schuster, New York	

In flight configuration of the Eurostar
communications satellite

(courtesy British Aerospace plc)

CHAPTER 2

Synchronous Orbits

2.1 INTRODUCTION

We have seen earlier that the time taken for a satellite
to orbit the earth, usually called the period of the
orbit, varies depending on the distance from the
satellite's path to the centre of the earth. The earth
itself rotates on its own axis and if the period of
rotation of the earth and the satellite's orbit are the
same, or in a simple ratio, the orbit is said to be
synchronous.

The satellite has to orbit the centre of the earth but
there is an infinite range of synchronous orbits of
different elliptical shapes, inclinations, period ratios,
and both ways of rotation. The inclination of an orbit
is the angle the plane of the orbit makes with the plane
of the equator.

In the special case of a circular orbit of zero inclination, orbit direction from East to West and a period
equal to a mean sidereal day the satellite appears
stationary to an observer on the earth's surface. This
orbit is the only one of real significance for direct-
to-home satellite broadcasting and is termed 'geo-
stationary' orbit.

Other synchronous orbits are used for satellite
communications to latitudes above 70° because of the low
elevation angles to geo-stationary satellites from such
locations. All orbits other than geo-stationary, of
course, suffer the considerable disadvantage of requiring
earthstations to track the satellite continually. In
addition, for many orbits, the satellite is not visible
all the time from all communication points and a
continuous service requires the provision of at least
two satellites in different phases of the orbit and
suitable changeover arrangements.

For example, highly elliptical orbits with the apogee
over the Northern Hemisphere, an inclination of about
63° and a 12 hour period have been used by the USSR for
satellite communications. This orbit, used by the
'Molniya' satellites, has an apogee of about 40,000 km
and perigee of about 1000 km.Such orbits may be unsuitable
for direct satellite broadcasting to very low cost
receivers.

The details of the geo-stationary orbit are:-

Orbital period	23hrs 56mins approx
Radius	42164 km
Height above earth's surface	35786 km (22236 st.miles)
Speed in orbit	3075 m/second

Strictly speaking the period of the orbit is not 24 hours
This is the period of the solar day, and results from a
combination of the earth's rotation on its axis and its
orbit round the sun. For geo-stationary operation the
satellite needs to have an orbit of period equal to the
sidereal day. The mean sidereal day is the average time
between successive passes of a given star across the

crosswires of a fixed telescope on the earth's surface and is the true period of earth's rotation.

2.2 GEO-STATIONARY GEOMETRY

The geo-stationary orbit is a concept of enormous value because it results in the ability to use earthstations with fixed antenna pointing. To make use of this, we need to calculate the pointing angles from the earth-station to the satellite. This chapter and the appendices provide the necessary information.

The geo-stationary orbit is quite a long way out from the earth's surface, being in fact about 5.6 earth radii distant from the sub-satellite point. This means that comparatively high latitudes can be illuminated by beams from the satellite without the angle of arrival at the earth's surface being too low. Low angles (say below 20°) are difficult to use for satellite broadcasting because many people would be in the 'radio shadow' of hills, trees and other buildings. Also, because of the long distance traversed through the atmosphere by low angle beams, they suffer extra attenuation from adverse weather conditions at the very high frequencies to be used.

In practice latitudes up to (and down to) at least 60°N (and 60° S) can be covered from the geo-stationary orbit provided the satellite is directly south or north of the coverage area. For good reasons, which will be discussed later, a broadcast satellite is usually placed up to 30° west of its target area. This lowers the angle of arrival of signals from the satellite to about 20° at latitudes 56°N (or S).

Angles of elevation and azimuth are used to describe the
pointing directions from an earthstation to the satellite.
The elevation angle is simply the angle the arriving
signals make with the local horizontal. Azimuth is
usually measured clockwise from true north, that is,
east of true north (ETN).

The equations used to calculate angles of elevation and
azimuth are derived in appendices.   When
calculating these angles it is necessary to work out the
distance from the earthstation to the satellite, usually
called the 'range'. The 'range' is useful in itself as
it can be used to work out the exact spreading of the
radio signals from the satellite due to the distance from
the coverage area.

The quantities needed to work out the range (d) and
angles are:-

   Longitude difference between earthstation λ
                    and sub-satellite point
   Latitude of earthstation ϕ
   Earth's equatorial radius (R)   6378.16 km
   Height of satellite above sub-satellite
                           point (h)   35786.3 km

the value of λ is positive if the satellite is east of
the earthstation.

The equations are:

$$(\text{range(d)})^2 = h^2 + 2R(R+h)(1-\cos\phi\cos\lambda)$$

$$\cos(\text{elevation}) = \frac{R+h}{d}\sqrt{(1-\cos^2\phi\cos^2\lambda)}$$

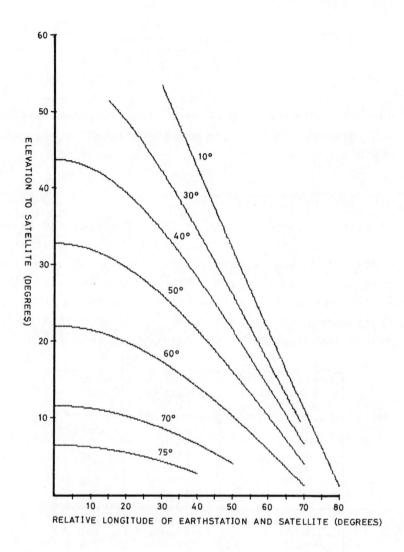

FIG.2.I ELEVATION ANGLE TO SATELLITE FOR DIFFERENT
EARTHSTATION LATITUDES

$$\sin (\text{azimuth}) = \frac{\sin \lambda}{\sqrt{(1-\cos^2\phi \cos^2 \lambda)}}$$

The relationships between elevation, longitude difference and earthstation latitude are illustrated in Fig.2.1. and Fig 2.2.

2.3 THE 'LINE' IN THE SKY

From any particular location, the elevation and azimuth angles to satellites in the geo-stationary orbit are in a fixed relation to each other. One way of illustrating this is to draw up a table of elevation and azimuth angles for satellites at different positions in the geo-stationary orbit as seen from a particular latitude. Such a table for $51^{\circ}N$, the latitude of London, is:-

Location minus Sub-satellite longitude	Elevation degrees	Azimuth degrees
70	3.7	254.2
60	9.8	254.8
50	15.5	236.9
40	20.7	227.2
30	25.2	216.6
20	28.6	205.1
10	30.8	192.8
0	31.6	180.0
-10	30.8	167.2
-20	28.6	154.9
-30	25.2	143.4
-40	20.7	132.8
-50	15.5	123.1
-60	9.8	114.2
-70	3.7	105.8

When the sub-satellite point is at the same longitude as the earthstation the satellite is at the orbital location which has the highest elevation from the earthstation and

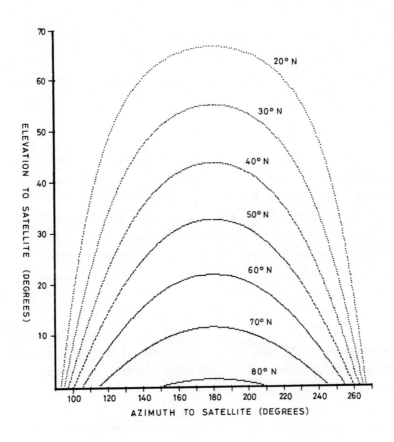

FIG.2.2 AZIMUTH AND ELEVATION OF THE GEO-
STATIONARY ORBIT AS SEEN FROM VARIOUS
LATITUDES

is due south from it (north in the Southern Hemisphere).
The maximum angle of elevation is, of course, wholly
dependent on the latitude of the earthstation, being 90°
for a location on the equator and falling to zero at
latitude   where:-

$$\cos \phi = \frac{R}{R + h} \quad ; \quad \phi = 81.3 \text{ degrees}$$

As the earthstation points towards satellites at
longitudes different from its own, the angle of elevation
drops and reaches zero at a longitude difference
where, for a latitude of   :-

$$\cos \lambda = \frac{R}{(R + h) \cos \phi}$$

This means, at latitude 51°, a satellite at 76.1° east
or west of the earthstation is on the horizon. This
theoretical limit of operation is only a geometric
result; in practice a satellite at 10° elevation
represents a limit of service for specially sited
communications earthstations and 20°C for home reception,
due to atmospheric absorption, screening, etc.

(Low angle reception is particularly important should
anyone on Europe seek to receive signals from the east
coast of North America (or vice versa)).

Fig 2.2 shows the azimuth and elevation for the geo-
stationary arc of satellites as seen from various
latitudes. Earthstations designed to work geo-stationary
satellites need only be able to point along these lines
to access different orbital positions. The design of
earthstations can take advantage of this fact and limit

the degrees of movement of the reflector or feed to
point at this 'line in the sky' (see also Fig 5.6).

2.4 LIFETIME IN GEO-STATIONARY ORBIT

At about 35,000 km from the earth's surface, satellites
in geo-stationary orbit do not suffer from wind and
weather! The atmosphere has, at that distance, become
negligible and the satellite is functioning in the near
perfect vacuum of space.

However, a number of forces act on the satellite to
disturb its equilibrium and deflect it from its
continuous circling of the earth. These forces,
technically referred to as 'perturbations', are counter-
acted by the satellite's propulsion system but,
eventually, succeed in forcing the spacecraft out of
orbit when the satellite's propellant is used up. The
satellite, which may still be functioning perfectly
electrically, has then to be switched off because its
uncontrollable path will interfere with other satellites.

'Dead' satellites continue to drift near the geo-
stationary orbit. As there is no atmosphere there is no
drag to decrease their velocity and cause them to descend
towards the earth's surface. There is, therefore, no
question of re-entry and possible collision with earth.
Eventually it may be necessary to remove space debris
and it may be possible and economic to refuel and
recommission spacecraft. However, considerable developments in space engineering are required before that point
is reached. The USA Space Transportation System (STS,
or space shuttle) is a first step, but at present it
only reaches low earth orbit (at about 300 km).

The most important perturbations are those due to the
gravitational fields of earth and sun. The geo-
stationary orbit is in the earth's equatorial plane but
this plane is inclined relative to the moon's orbital
plane round the earth and the earth's orbital plane
round the sun. The effect on the satellite's orbit is
complicated depending on the time relation between the
orbit and the sun and moon positions. However, the
effect of both these gravitational fields is to pull
the satellite out of the equatorial plane and cause its
orbit to become inclined. The magnitude of the effect
varies over the sun/moon cycle and is about 1 degree
per year.

Early communications satellites did not attempt to
correct this perturbation to the orbit (which produces
an apparent figure-of-eight motion as seen from the
earth's surface). As the inclination of the orbit
builds up, the satellite swings further north and south
of the equator producing an ever larger oscillation.
However, tracking the satellite during its daily figure-
of-eight pattern poses no problems for a large communications earthstation. Satellites designed for use with
simple earthstations set on fixed bearings must be kept
in the equatorial plane and the orbital adjustments
necessary to achieve this are called 'north and south
station keeping'. North and south station keeping
requires velocity changes amounting to about 50 m/s
per year which is provided by the satellite's propulsion
system. Direct broadcasting satellites are all equipped
with sufficient fuel for north and south station keeping
throughout their lifetime.

The second largest orbital perturbation is caused by the
lack of symmetry in the earth itself. Although our
earlier geometric calculations on elevation and azimuth

angles assumed the earth to be a perfect sphere, this
is not the case. The departure from symmetry only
affects the most precise calculations of earthstation
pointing, but it is important in any discussion of
satellite orbital motion. Since Newton's time it has
been known that the earth bulges at the equator and is
flatter towards the poles. This shape is called an
'oblate sphere' and affects orbits which are inclined
to the equatorial plane. As we have seen the geo-
stationary orbit is of zero inclination and is not
affected. Of course, some of the other
inclined orbits briefly mentioned earlier have to take
account of this effect.

However, in addition to being 'oblate' the shape of the
earth is such that the equator is not an exact circle.
This lack of symmetry does affect the geo-stationary
orbit by creating a component of the gravitational field
in the direction of the satellite's motion causing it to
speed up or slow down and drift east or west of the
assigned station. Depending on the required orbital
station's relation to the equatorial shape, east and
west station keeping requires the expenditure of fuel to
give velocity changes of up to 2 m/s per year. All
geo-stationary communications and broadcast satellites
are kept on station in the east/west direction, as to
drift off-station would mean that the satellite would
move out of sight of the points on the earth's surface
it is connecting together.

In addition to gravitational effects, satellite orbits
are also affected by forces arising from the earth's
magnetic field, solar pressure and micro-meteorite
bombardment. The most important of these is solar
pressure, although, unless the satellite is of much
lower mass and larger physical dimensions than currently

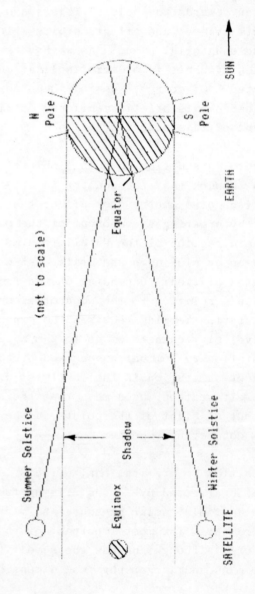

Fig.2.3 Solar Eclipse of a Geostationary Satellite

designed or planned broadcast satellites, it has a
negligible effect on the satellite's orbital motion.

Satellites now in orbit have been designed to last up to
7 years and are loaded, before launch, with sufficient
propellant to counteract the calculated perturbations
for that period. Allowance has to be made in addition
for the propellant required to position the satellite
in its assigned geo-stationary position following launch
and for any repositioning of the satellite which may be
required during its lifetime. Satellites now being
designed are planned to have 10 years life. It is
possible that this represents the maximum for inaccess-
ible spacecraft as the extent of technology development
over 10 years makes a generation of satellites obsolete.

It is possible to use the radiation pressure on the
solar panels to assist attitude control. "Sailing" in
this manner saves some of the long term fuel store.

2.5 SOLAR ECLIPSE AND BROADCASTING SATELLITES

We have seen earlier that the best position in the
geo-stationary orbit for a communications satellite is
at as close a longitude as possible to the earthstations
it serves. This policy results in the maximum possible
elevation angle to the satellite from the earthstations
giving least attenuation through the atmosphere and
least risk of blockage by buildings or hills. For
broadcasting satellites, however, a very important
additional factor has to be taken into account: solar
eclipse.

Figs.2.3 and 2.4 show the solar system with the earth
and a geo-stationary satellite at different times of
year. Because the equatorial plane, and hence the plane

38

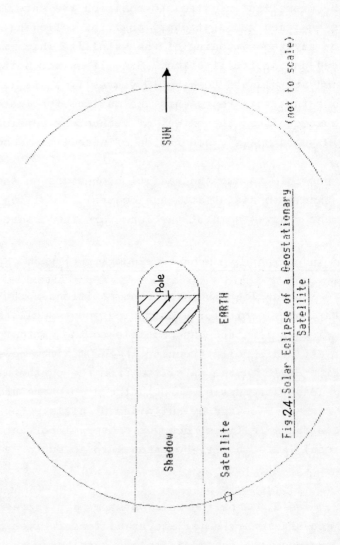

Fig.24. Solar Eclipse of a Geostationary Satellite (not to scale)

of the orbit of the satellite, are inclined to the
earth's orbital plane, the sun, earth and satellite are
only directly in line at the two equinoxes. At other
times they are never completely in line and, for example
at the winter solstice, the satellite is above the
earth's orbital plane when it is nearer the sun than
the earth is and below the plane when it is further from
the sun. It can be seen that during the two equinoxes
the earth can eclipse the satellite, coming between
satellite and sun and cutting off the satellite's solar
energy supply. The duration of the eclipse at the
equinoxes is determined by the sizes of the earth and
sun, and the distances from earth to sun and satellite.
The same constants determine the relation between date
and eclipse duration and the curve in Fig.2.5 holds for
all geo-stationary satellites.

Although modern communications satellites carry
rechargeable batteries to power the satellite during
eclipses this is simply not practical at present for
broadcasting satellites. The mass of the batteries
required to supply the power demands of the high power
amplifiers on board communications satellites is a very
significant part of the total mass. The power required
by these amplifiers is between 15 and 50 watts each, but
the broadcast satellites require ten times this power.
Limited arrangements may be possible in specific cases
to keep one or two channels on during eclipse but the
cost in mass terms is, in general, too great.

The solution to this problem is to limit the broadcasting hours to times during which the satellite is in
sunshine. Fortunately the satellite orbit is such that,
at worst, sun power is lost no earlier than about 36
minutes before "midnight". Moreover the midnight
referred to is the midnight at the satellite's longitude

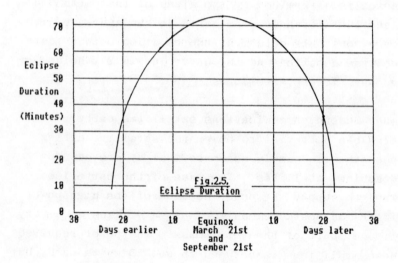

Fig.2.5.
Eclipse Duration

and, provided elevation angles of earthstations in the
target area allow it, the satellite can be positioned to
the west of its service area to push the onset of
eclipse apparently later by up to about 2 hours.

For this reason the broadcasting satellite location for
the UK is at 31°W serving an area roughly 30° to its
east. The same considerations have been used in drawing
up the plans for broadcasting satellites in all parts of
the world. However, where the service areas are far to
the north or south and elevation angles to the satellites
are already low there is less scope for using westward
positions. For example the Scandinavian countries have
agreed to use 5°E as their broadcast location even though
parts of the service are close to that longitude. The
Scandinavians can be consoled by the fact that there are
no eclipses during winter, which is the peak season for
television.

Chapter 4 gives the orbital locations and other details
of some of the broadcasting satellite systems planned by
the Administrative Conferences in 1977 and 1983 (WARC
and RARC).

2.6 <u>SUN OUTAGE</u>

During the equinox seasons (as the opposite condition to
satellite eclipse which occurs about 12 hours later) the
sun, satellite and earth are all in line. The satellite,
of course, is not big enough to cast a shadow on the
earth's surface but the effect called sun "outage" occurs.
A more informative description of the effect is sun
blinding, and the problem arises because receiving
earthstations which are pointing at the satellite are
also pointing directly at the sun for a short period.
The sun, being a hot radiating body, generates random

electromagnetic signals and these are more powerful than
the wanted satellite signals. The result is that, for
a short time, the television signals radiated by the
satellite are affected by random noise signals. The
duration of the effect depends on the size of the earthstation's antenna and is typically 10 minutes for a 0.9
metre antenna on the most affected day. The worst day,
of course, occurs twice a year once during each equinox
(and shorter period one day either side of it).

In total 8 days a
year may be affected at any one place. The exact time
of day and date of longest outage depends upon the
position of the earthstation for any particular
satellite and are totally predictable.

There is no way of avoiding sun outage from a particular
satellite and communications satellite systems requiring
continuous service have to use two earthstation antennae
and access two satellites at different longitudes.
Broadcasters will have to take sun outage into account
in programme schedules to avoid the frustration to
viewers caused by missing the end of an exciting film or
the implications of losing a commercial in some parts of
the service area.

(The degree of impairment likely to occur is serious but
the required signal may not be totally obliterated
during these periods. See Appendix).

2.7 LAUNCH SYSTEMS

2.7.1. Geostationary satellites

Artificial earth satellites have been made possible by
advances in rocket technology and a study of broadcast
satellites would not be complete without at least an

appreciation of launcher technology.

Two basic methods will shortly be available for commercial satellite launches. The first is to use one of the many expendable rocket launcher systems already available including mature systems such as Thor-Delta and Atlas-Centaur and the newer European Space Agency's Ariane system. The second method is to employ the USA Space Shuttle which represents the first re-usable vehicle launch system and which will shortly be available for commercial launches. Both expendable and re-usable launch vehicles depend upon the basic rocket theory.

Rockets obtain their acceleration by ejecting a high velocity stream of material in the direction opposite to motion. The thrust produced is more the greater the rate of mass ejected by the rocket and the greater its velocity. Although the thrust of the rocket is essentially constant during the 'burn', the mass of the rocket continually gets less as rocket fuel is expended. As a result acceleration is continually increasing until fuel burn-out and the relation between velocity and time is exponential. Fig.2.6 illustrates the fundamental relationship leading to the basic rocket equation which is derived in the appendices:

$$V = V_O + I.g.\log_e \frac{M}{m}$$

where:-

V = velocity at any time

V_O = starting velocity (equals zero for first stage)

I = specific impulse of rocket fuel (see later)

M = starting mass of rocket and fuel

m = mass of rocket and remaining fuel

g = acceleration due to gravity

44

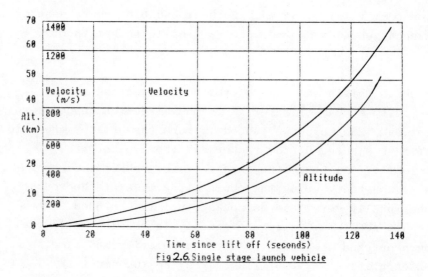

Fig 2.6. Single stage launch vehicle

This equation shows that the velocity change achieved by
a rocket is determined by the mass of the vehicle
(including the payload), the mass of fuel it carries and
and rocket fuel's potency. For the special case in which
all the fuel available is burnt then the velocity change
achieved is a maximum for the rocket.

The maximum velocity change possible using a particular
fuel is determined by the mass ratios of structure, payload and fuel. These factors are determined by the
strength to weight ratio of the materials used in the
rocket's construction and the ingenuity of the designers.
The 'dry mass', that is the mass of the rocket and payload, is made as low as possible for a given quantity of
fuel and thrust level by good mechanical design, avoiding
oversizing structural components and by using new
technology materials like carbon fibre. Of course,
particularly in expendable launchers, the cost of
materials and construction is also an important consideration.

Astonishing progress has been made in the design of
rockets and, for example, at lift off, the European
Space Agency's Ariane launch vehicle has an all-up mass
of 207 metric tons of which only 26 metric tons is taken
up by structure, engines, tanks and controls. If the
same mass ratio were applied to a small family car of
1200 kg all-up weight than the empty vehicle would weigh
150 kg (330 lb) and could be carried by two men. (However such a car might easily be blown over in a strong
wind and the road-holding might not be very good).

The specific impulse is a property of the rocket's fuel
and is related to the amount of chemical energy released
when the reaction is set off and to the mass of the fuel
itself. The specific impulse is defined by the equation:

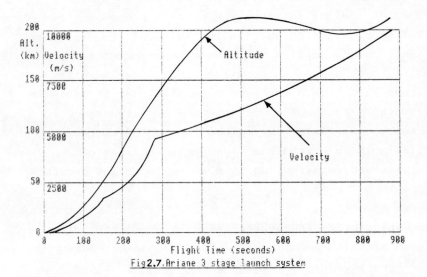

Fig **2.7**. Ariane 3 stage launch system

$$\text{Specific Impulse} = \frac{\text{Thrust force}}{\text{Weight/second of fuel used}}$$

The greater the thrust force produced for the consumption of a given weight of fuel per second, the greater the specific impulse. Obviously if two fuels produce the same thrust but the weight of one used per second is lower, then that fuel has the higher specific impulse. One of the most common rocket propellants makes use of liquid hydrogen and liquid oxygen. Although they must be stored and loaded at extremely low temperatures (hydrogen -253°C, oxygen -183°C), their high specific impulses, non-toxic nature and ready availability makes it worthwhile to overcome the cryogenic difficulties. Common liquid and solid propellants are:-

Fuel	Oxidiser	Specific Impulse (mS^{-1})
Liquid hydrogen	Liquid oxygen	430
Kerosene	Liquid oxygen	328
Di-methyl hydrazine	Di-nitrogen tetroxide	309
Hydrazine (mono-propellant)		225
Solid propellants		295 - 325

2.7.2. Multi-stage rockets

The attainment of geo-stationary orbit requires a combination of velocity and altitude beyond the capability of a single rocket stage with currently known fuels and materials. The reason for this is clear from the earlier discussion on the relation between mass ratios, specific impulse and velocity change. It should, however, be noted that the rocket equation quoted earlier does not take account of gravitational energy changes

which have, of course, to be allowed for in determining
the energy needed to put a satellite into orbit.

The multi-stage launcher typically uses three stages to
launch a satellite into the elliptical transfer orbit,
or Hohmann's ellipse, described earlier. The mass ratios
of payload (plus dry mass) to fuel for each stage are
similar and because each stage forms the payload for the
one below, the masses of each are in a geometric
progression. For example the masses of the Ariane
launcher are:-

	Payload Mass p (kg)	Stage Mass s (kg)	Fuel Mass f (kg)	Ratio f/(p+s+f)
Stage 3	2000	1200	8200	0.72
Stage 2	11400	4400	34000	0.68
Stage 1	49800	20000	145000	0.68

In a multi-stage system each stage contributes a similar
increase in energy, both kinetic and potential, to the
final payload and then separates from the rocket system.
Small retro-rockets are fitted to stages one and two and,
after the securing system is released, these are fired to
retard the burnt-out stage so that the next stage may
draw ahead and suffer no disturbance when ignition occurs.
After burn-out the first stage follows a ballistic
trajectory and comes down up range. For this reason a
considerable length of clear ocean or uninhabited land
is needed for a launching range.

The Space Transportation System or Space Shuttle is
designed to re-use as much as possible of the system to
minimise launch costs. However, for the reasons given
earlier, it must use some element of staging. The
arrangement is to use solid 'strap-on' booster rockets
which are ignited at lift-off with the main engines but

burn out and drop away after a few seconds. In addition
the main fuel tank, on which the Shuttle appears to ride,
is jettisoned after a few minutes, but it is equipped
with a parachute system and is recovered for reconditioning
and re-use. After separation of the boosters and the
tank the Shuttle itself becomes the upper stage carrying
payloads to a low earth orbit of 296 km.

Satellites released from the Shuttle in low circular
earth orbit require two further rocket stages to attain
geo-stationary orbit. The first of these is used to put
the satellite into the transfer orbit and, since it is
fired at the lowest altitude of this orbit, is called
the Perigee stage. The circularisation of the transfer
orbit is achieved by increasing the velocity at apogee
by firing the apogee boost motor.

Satellites released from Ariane and other expendable
rocket systems are normally in transfer orbit and are
only required to carry the apogee boost motor to achieve
a geo-stationary orbit.

2.7.3. Launch systems and costs

Intense competition is building up between the USA's
Space Transportation System (Space Shuttle) and the
expendable launch systems. The USA, Europe, Japan and
the Soviet Union all have expendable launch systems.
The capacity of the Shuttle is likely to be adequate for
the largest single launch satellites which will ever be
required, larger space structures probably being launched
in sections and assembled in orbit. To accommodate a mix
of loads the Shuttle can carry satellites and other
payloads together into low circular earth orbit up to its
maximum carrying capacity of 30,000 kg. The Shuttle

charges users by mass or by length of cargo bay used,
whichever is the most demanding (see Chapter 7).

Satellites designed only for a Shuttle launch can be
optimised by adjusting their shape so that the two
costing criteria are met, as nearly as possible,
simultaneously. In making cost comparisons, it has to be
remembered that an extra perigee stage is required for
Shuttle launches.

Some expendable systems can also carry more than one
satellite into transfer orbit. Ariane, for example, can
carry two satellites of combined mass slightly less than
the total carrying capacity. The extra mass is required
for the adaptor system which transfers the thrust of
launch to the upper passenger.

Further details of the cost of launch systems are to be
found in Chapter 7.

2.8 THE SPACE SEGMENT

Up to this point the discussion of the satellite system
has been concerned with matters of principle. The
practical implementation of these principles is a
complex subject if every aspect is covered.

Although it is beyond the capacity of this book to cover
all these details the remainder of this chapter will
outline the main features that influence the cost of the
project and so form the basis for further discussion in
Chapter 7.

2.8.1 <u>Single Satellite or Cluster?</u>

A single communications satellite in geostationary orbit
is the least 'space segment' possible. However, even
one spacecraft in orbit represents a significant, and
perhaps essential, communications facility and consideration has to be given to what happens to the service if
it fails completely. Examination of this problem poses
another question: what is the best way of providing a
continuous service having regard to the finite,
predictable, lifetime of the satellite? As will be
seen later, a spare satellite (or satellites) is often
considered necessary and then it is worthwhile seeing
whether the spare satellite(s) can earn its living in
orbit rather than just remain on-station, consuming
station keeping fuel, without contributing to the
earnings of the system. Two or more operational
satellites in orbit provide a system which is immune to
complete breakdown if one satellite fails, and this is
called a 'cluster'. From the earth, a cluster appears
as a single satellite and the earthstations do not need
to know which member of the cluster is carrying the
service for any given application.

A cluster of satellites forms a way of providing a total
capability in excess of the abilities of a single
satellite. The members of the cluster are designed to
work together as a system and are planned in a
complementary manner. There are various ways of dividing
the total task among the members of the cluster but the
most effective way, for Direct Broadcast applications,
is to use frequency division. Using this technique,
each member of the cluster is self-contained and receives
and re-transmits some of the assigned broadcast channels.

The cluster is kept within the 'box' assigned to the orbital station, say 0.1 N-S and E-W, by joint station keeping manoevres. Each satellite in a cluster must be able to transmit on channels which are different from the other members but must also carry spare frequencies.

The members of a cluster of a Direct Broadcasting system are all well within the beamwidth of even a comparatively large (say 3m diameter) receiving earthstation. The same is not necessarily true for the transmitting earth-station. This may have to be of such a size that the beamwidth from it is unable simultaneously to illuminate members of the cluster at opposite 'ends' of the 'box'.

The controlling earthstation may suffer from the same disadvantage of requiring an antenna for each member of the cluster. Telemetry and control frequencies have to be unique for each member of the cluster but this is not a significant difficulty.

Clusters provide a way of engineering large 'virtual' satellites, and may limit the number of spare spacecraft required. However, they have not been used for point-to-point applications. In such applications, a spare satellite, identical in all respects to the 'workers', can be located in an adjacent orbital location and used to earn revenue by being employed for applications which can be interrupted in emergency. Such applications are termed 'pre-emptible' and, of course, attract a lower rental than fully protected channels. Direct broadcast applications, however, may favour clusters because a prescribed orbital location must be used for all the transmissions.

The main disadvantage of a cluster of small satellites working together to provide both main and reserve capacity is cost. The alternative solution of launching primary and spare satellites of larger dimensions is usually cheaper, especially if pre-emptible services can be sold. Apart from all the fixed costs which do not vary with satellite size, a number of important on-board units, for example the antennae, have to be provided on each spacecraft irrespective of how many channels it provides. The small satellites in a cluster may be relatively expensive.

Apart from the effect of technical complexity on the reliability of a cluster satellite, the laws of probability do not favour division of a total system into small units. An illustration is to consider a simple system provided in one of two ways:

1. Two satellites each equipped with one main (M) and one reserve (R) transmitter
2. One satellite, of twice the capacity, equipped with two main and two reserve transmitters

Although the same number of transmitters are in orbit, providing suitable switching is used, the second arrangement can continue with no loss of service even if two transmitters fail. In the first case, however, of the six ways in which two transmitters can fail, two of them cause complete loss of one service. As the arrangements have the same resistance to single or triple failures, arrangement 2 offers the better reliability. A summary of the failure modes and the effects on the service are:

54

Failure mode	Arrangement 1	Services Lost	Arrangement 2	Services Lost
No failure	M1 R1 M2 R2	–	M M R R	–
Single failures	R1 M2 R2	–	M R R	–
	M1 M2 R2	–	M R R	–
	M1 R1 R2	–	M M R	–
	M1 R1 M2	–	M M R	–
Double failures	M2 R2	1	R R	–
	R1 R2	–	M R	–
	R1 M2	–	M R	–
	M1 R2	–	M R	–
	M1 M2	–	M M	–
	M1 R1	1	M M	–
Triple failures	R2	1	R	1
	M2	1	R	1
	R1	1	M	1
	M1	1	M	1
Total failure		2		2

However, it is necessary to go into much greater detail
on the reliability of the component parts before a final
decision can be taken on this subject.

2.8.2. Single and Multiple Missions

In the early days of satellite communications the most
that could be achieved in geostationary orbit was one
electronic communications assembly receiving, amplifying,
frequency changing and re-transmitting one signal from
earth. The electronic assembly used to do this
is termed a transponder. As capabilities developed
several signals were first passed through the one
transponder, then several transponders working on
adjacent frequencies were provided. Now the possibility
exists of carrying assemblies of transponders, each
assembly working on a different communications task.
Such a satellite is termed multi-role. In addition to
satellites peforming different roles, systems can also
be designed to carry out similar missions on several
different frequency bands simultaneously. These are
sometimes called hybrid systems. The common parts of
the satellite not concerned directly with the transponder
functions are often called the platform. The transponders and antennae are referred to together as the
payload.

It is obvious that a multi-role or hybrid mission is
only possible if an orbital location can be found at
which all the frequencies on board the satellite can be
used. Such an orbital position may have to be a
compromise for a multi-role system. For example, as
discussed in Chapter 1, an orbital location to the west
of the target area is required for direct broadcasting
because of eclipse problems. However, a point-to-point
mission, sharing a satellite with direct broadcasting
transponders, would continue to be available for use
during eclipse and an orbital location as close as
possible in longitude to the target service area would
be preferable to maximise coverage at high latitudes.
As well as the uplink and downlink frequencies being
clear for use from earth to satellite and vice-versa
the downlink of each part of the multi-role system must
not interfere with the uplink of the other. As the
downlink powers are very much greater that the power of
the uplink signals received at the satellite, great care
must be taken with filtering and screening within the
satellite's payload.

As for orbital position, the station keeping accuracy
may be a compromise between the various missions of a
multi-role satellite. It is unlikely to be a serious
problem but, nevertheless, it is a factor to be
considered if one mission requires very much tighter
station keeping accuracy than the others. The amount of
station keeping fuel expended depends both upon the
accuracy requirements and the mass of the satellite.
Clearly it would be a waste of resources if a massive
payload requiring low accuracy were to be kept on
station to a high accuracy because of the requirements
of a small transponder sharing the platform.

All the various missions of a multi-role satellite are launched together and any spare satellites are shared by the system as a whole. Therefore the roles must be compatible in terms of the reliability to be achieved and the replenishment policy to be adopted. Even if such compatibility exists in principle there are considerable problems to be solved in practice in sharing a platform. If, for example, the transponders of one of the missions fail in orbit the spare satellite will have to be brought into service. The good transponders on the first satellite may then have to be prematurely retired. If costs have to be apportioned between the various missions then all these operational problems have to be addressed along with the obvious factors of mass and power requirements of the component sets of transponders.

2.8.3. Spares and Replenishment

The need for spare equipment on board satellites, spare satellites in orbit and the provision for continuity of the space facilities are all related topics. Spacecraft fail and wear out in a number of ways both because of inherent lifetimes, outside influences and random catastrophic failures. Any failure may affect all or only part of the satellite and a discussion of all this is included later in this chapter. For the moment we will consider only the effects of partial or total satellite loss on the overall system.

The simplest satellite systems employ one satellite in
operation, one spare satellite in an adjacent orbital
position and one ready for launch and in storage on the
ground. In some cases the in-orbit spare satellite may
be switched on and used for pre-emptible traffic as
described earlier. The ground spare need not be fully
completed as the longest 'lead-time' to a new launch is
often the availability of the launch vehicle itself.

In the event of a total failure of the duty satellite in
this simple system, the spare satellite is brought into
operational service and the earthstations re-point to
its position. Preparations are started immediately for
the launch of the ground spare and, perhaps, the
purchase of a new ground spare to restore the system.
For broadcasting satellites, the spare satellite has to
be at the same orbital location as the working spacecraft.
Repointing of millions of receive-only earthstations
would otherwise be required. The broadcasting plan
also requires rigid adherence of each country's transmissions to a given orbital location for reasons of non-
interference. Systems with more than one operational
satellite may share a common in-orbit spare. The method
of bringing such a shared spare into use is the same as
for the case of a single spare system described.

The station keeping fuel for in-orbit spare satellites
not used for pre-emptible traffic may be conserved by
placing them initially in an inclined orbit and omitting
North-South station keeping. As described in Chapter 1,
perturbations affect the inclination of geostationary
orbits and, by careful analysis, an initially inclined
orbit can be achieved which will 'decay' to a zero-
inclination orbit a known time after orbit injection.
By this means the spare satellite can be arranged to be

in the zero-inclination orbit at the time when it is most likely to be required for operation. Once that orbit is reached, North-South station keeping is commenced to prevent inclination building up in the opposite direction. If the spare satellite is required earlier than anticipated then station keeping fuel has to be expended to reach the required orbit for operation and the benefits of the approach are partially nullified. The worst situation would occur if the spare were needed immediately after it was launched due to a coincidental failure of the primary satellite. The available station keeping fuel of the spare would then be substantially less than if it had been injected into a nominally zero-inclination orbit and manoevreing fuel only used to correct for residual errors in the apogee motor firing.

Complete failure in-orbit is fortunately rare and the more usual situation encountered is that of a partially failed satellite able only to carry out part of its assigned mission. The decision to activate the spare satellite can be operated from an adjacent orbital location. In this case the partially failed satellite can continue to function with impaired performance and share the load with the spare. If, as is the case with direct broadcasting systems, the spare satellite has to operate from the same orbital location and using the same frequencies as the partially working spacecraft, then a complete takeover is necessary. There is little experience of the critical situation requiring the balancing of the loss of service due to premature 'ageing' of the satellite system with the alternative cost of purchasing and launching further spacecraft.

Designing a 'replenishment' strategy to keep the requisite number of transponders operating in orbit is

a specialised subject. It involves both engineering
reliability and financial factors. In the case of direct
broadcasting satellites for which the audience may be
expected to build up over several years, the optimum
approach might be to delay the launch of the spare
satellite until the importance of the service justifies
expenditure of life-time of the spare. The delay in
restoring service in the event of a primary satellite
failure would then be determined by the time needed to
bring the stored spare satellite to launch readiness and
to arrange for a launch. This time might be 4 to 6
months. Unfortunately, as a result of the rigid planning
necessarily adopted by the WARC 1977 plan each broadcast
satellite has a unique orbital position and broadcast
frequency. This means that there is only a limited
opportunity for a number of countries to share a common
spare spacecraft.

2.8.4. System Reliability

The first thing to consider is the reliability of an
individual communications satellite. Experience confirms
as expected that the launch and deployment phases of the
satellite's life are the times when it is most at risk.
The main hazard is, of course, launch vehicle failure
and a substantial number of carefully prepared
communications satellites end up as debris in the sea
following the sending of a 'destruct' signal to the
rocket. The safety of the launch site and surrounding
area to say nothing of the possibility of an off-course
satellite launcher becoming, inadvertently, a ballistic
missile, requires the launch safety officer to destroy
any vehicle deviating significantly from the expected
performance.

Following launch into the elliptical transfer orbit, the apogee boost motor is fired. Satellites have been lost during this phase of the mission because of mal-operation of the motor. The solid-propellant rockets often used are uncontrollable once fired and the controlling earth station can only ensure that the satellite is correctly orientated and that the firing signal is given at the correct instant. The newly available liquid-propellant apogee boost motors will allow earthstation intervention in the process of orbit circularisation and may help to lower the risk of the process.

The deployment of solar arrays and antennae is required at the start of the operational life of a modern communications satellite. Solar arrays have to be sized to generate sufficient power for the satellite's equipment. They are steadily improving in efficiency but spacecraft power requirements are increasing and it is necessary to 'pack' the array up in some manner to allow its considerable area to be stowed in the confined space available for payload in the launch vehicle. Similarly the antennae are often too large for the launch vehicle because their dimensions are determined by the beamwidth required of the radio signals from the satellite and the wavelength employed. Deployed antennae are therefore common. Deployment of structures in space is achieved by a variety of means but they all present a potential hazard to the mission.

The technical assessment of realiability and practical experience are combined when obtaining an insurance quotation. The cost of insurance is a not insignificant part of the system cost.

2.9. SATELLITE PLATFORM

2.9.1. Spin & Body Stabilisation

The motion of a geostationary satellite around the earth
was described in Chapter 1. If the situation is
considered from the satellite's view-point the earth
appears to rotate around the satellite. This view is
an exact analogy to the pre-Copernican view mankind took
of the earth and sun. However, as on the solar scale,
the satellite-centred view is useful when considering
some situations.

It is useful, for example, when describing the attitude
control needed for a geostationary communications
satellite. A satellite training a tight pencil of radio
waves at the earth needs a control system to keep the
beam on target. A direct broadcasting satellite requires
a very advanced form of beam pointing as the tolerance
in direction is only $\pm$ 0.1. The basic need for pointing
has had an important influence on the overall design of
communications satellites.

As the earth appears to rotate around the satellite the
antennae must be turned to face the coverage area. The
satellite platform needs to be stable in space and
requires 'stiffness' in rotational motion. This inertia
or 'stiffness' in rotation is provided by a gyroscope
which 'resists' being turned in space about any axis
other than that of its own rotation. All the early
designs used the satellite itself as a gyroscope or top
and caused it to spin about an axis. Although this
design is still widely used, an alternative arrangement
is slowly taking over; that is one in which the satellite
itself does not spin but contains spinning gyroscopes.
Such a satellite is termed "body stabilised".

The "spin-stabilised" or rotating satellite is shaped like a top with the solar cells arranged on the outer surface of the cylinder. The cylinder spins at about 50 r.p.m. about its long axis (which is parallel to the earth's axis when the satellite is orientated for operation). The antennæ are mounted from the top (or north) face and the system to which they are fixed is driven by electric motors exactly cancelling out the platform's spin. This process is called, not unnaturally, de-spinning.

The body stabilised satellite is box shaped with the solar cells mounted on two 'paddles' fixed to the north and south faces of the satellite. The satellite rotates once per day so that one of the faces of the box always faces the earth. The booms supporting the solar arrays are motor driven at about one revolution per day to counteract the satellite's rotation so that the solar cells always face the sun. Attitude stability is provided by internal gyroscopes.

Comparisons between the two basic types of design are given in a tabular summary on the next page. Some of the areas of comparison are covered in more detail in the next section of this chapter.

	Spin Stabilised	Body Stabilised
Solar Array	Part of array only illuminated at any instant	Whole array illuminated continuously. Deployment required.
De-Spin	Antennae and transponders de-spun at 50 r.p.m.	Solar arrays de-spun at 1 revolution/day.
Thermal Radiator	Radiator available on annulus at expense of solar cells	Space for radiators on north and south faces
Thermal Balance	Inherently balanced throughout the satellite	Balance needs care
Antenna Mounting	Limited antenna space usually requiring deployment when in orbit.	Earth facing platform usable without deployment. East and west faces available with deployment.
Attitude Control	Satellite forms gyro. De-spin synchronisation required.	Internal gyros required. Control essentially not time critical.
Propulsion	Thrusters have to be pulsed. Can act as spares for each other.	Thrusters can be operated continuously. Spares needed.
Launch Sequence	Spin-stabilisation in transfer orbit and on station.	Spin-stabilisation in transfer orbit, de-spin and 3-axis acquisition on station.

2.9.2. Power System

The power system for broadcast satellites uses solar-
generated electricity. The solar-cells normally used
are about 10% efficient at the end of their life, but
improved cells are slowly becoming available giving 12-
15% efficiencies. The satellite carries re-chargeable
batteries to supply the essential 'platform functions'
during the periods of eclipse described in Chapter 2.
Specially designed Nickel-Cadmium batteries are used at
present and these provide storage of about 3 watt hours
per kilogram. Newer designs of battery are now in space
using Metal-Hydrogen couples and providing storage of
up to 9 watt hours per kilogram. The control of depth
of discharge of the batteries must be rigid as their
ability to survive for the lifetime of the spacecraft
depends upon this. The regulation of the power supplies
to provide a constant energy source for the platform and
parts of the payload in spite of eclipse, temperature
variations of the solar cells during the eclipse periods,
annual variations in solar energy received and steady
decay of the solar array efficiency is itself a highly
specialised branch of space engineering.

Appendix 4 shows that the effective area of the solar
array of a spinning satellite is less than 1/3 of the
actual area. This comes about because all the array is
not facing the sun at any time. Half of it is in
shadow and all of it, except for a narrow
strip, receives sunlight at other than the most
effective angle - a right angle. There is a compensa-
tion in that the array runs cooler than if it was in
sunlight the whole time and this improves the
efficiency of conversion of sunlight to electricity.
However, the nett result is still an efficiency of only
about 35% of a flat array turned to track the sun. The

solar array for body-stabilised satellites requires
deployment. Direct broadcast satellites, being
dependent on large amounts of solar power (up to 5 kW)
will probably all use body stabilised satellite designs.

A Bearing And Power Transfer Assembly (BAPTA) is used
in a spin-stabilised satellite to allow the whole
payload, comprising transponders and antennae, to be
de-spun. The power from the solar cells as to be fed
via slip-rings to the payload as does telemetry and
command data. Failure of the bearing causes loss of the
whole satellite and duplication of this element is not
easily possible. BAPTAs are also required for a body
stabilised satellite, though the rotational speed is
nearly 5 orders of magnitude slower. Two BAPTAs are
used, one for each solar array, and they are usually
driven by dual stepping motors. Failure of one BAPTA
to rotate would have a very severe effect on the
operation of the satellite as that solar array would
then fail to track the sun during the diurnal rotation.

The comparison of overall efficiencies of the power
generation for the two designs of satellite may be made
by estimating the mass required for a given steady
end-of-life power. Figures of about 1.5 watts/kilogram
are typical for a spin stabilised satellite. The body-
stabilised design achieves equivalent figures of about
2.3 watts/kilogram at the 500 watt level and up to
4.5 watts/kilogram at 2 kW. The larger the total power
requirement the greater the efficiency of the body-
stabilised solar array system.

2.9.3. Thermal Control

About 2/3rds of the electricity supplied from the solar arrays of a satellite is dissipated within the satellite as heat. It is therefore very important to be able to get rid of heat from within the areas of the spacecraft containing the amplifiers and other electronic components producing it. Unless this is done high temperatures may eventually lead to premature component failures. When switched off, equipment must not fall below its survival temperature.

Heat transfer within the satellite can only be by radiation or conduction when in geostationary orbit. There is, of course, no air to provide convection cooling. (Even if there was air inside the satellite there is no nett gravitational effect in geostationary orbit and therefore convection currents would not occur). Careful design and testing is needed to ensure that the satellite can radiate to space both the heat generated internally and that incident from sunlight to keep the overall temperature constant.

The spin-stabilised satellite is short of surface areas not facing the sun. The north face is taken up with antennae, the south with the apogee boost motor and, in any case, most of the internally generated heat arises from the de-spun electronics below the north face. The only surface available is that of the main cylinder which itself collects solar radiation. The normal arrangement is to include a radiator band near the centre of the satellite coupled, by internal radiation, to the de-spun section.

For the body stabilised spacecraft the heat producing assemblies can be mounted directly on the north and

south walls as these always face deep space and radiate
heat freely. In both designs of satellite the surface
finishes are chosen to give the desired absorptivity
and emissivity on the outer surfaces.

During eclipse some of the payload is switched off to
limit the battery requirements; the whole of it in the
case of many direct broadcasting satellites. The
thermal balance of the satellite is then very different
from normal and electric heaters may have to be used in
some places to prevent dangerously low temperatures
occurring. In normal operation the balance between
parts of the satellite is also very important to ensure
there are no 'hot spots'. As mentioned earlier no
convective thermal distribution can occur and high power
payloads may require 'heat pipes' to carry heat away
from operating equipment to cold surfaces. (A 'heat
pipe' is a sealed pipe containing vapour and liquid
in equilibrium and which is able to transfer heat very
efficiently by making use of the latent heat of
vapourisation and condensation of its contents). Some
high-power amplifier designers now make tubes in which
the main heat producing electrode, the collector, can
be mounted outside the spacecraft and directly radiate
heat to space.

2.9.4. Structure and Antenna Mounting

The structure of a satellite has to withstand the
acceleration and vibration of launch and remain stable
throughout the thermal cycling of years of eclipses in
orbit. Above all it must be light, as every kilogram
of structural mass reduces the payload capacity of the
system. The near constant weight of this structure
gives an advantage to larger satellites. The structure
of a medium sized satellite, say about 500 kgm,

will be about 20% of the total mass of the spacecraft,
whereas that for a larger satellite of 1300 kgm will be
about 12%.

It is obvious that the body stabilised satellite offers
more accommodation for antennae and the avoidance of
deployment mechanisms for the simpler missions. Many
of the present body-stabilised communications satellites
now in orbit use directional antennae mounted only on
the earth-facing side of the satellite and to not
require deployment. Missions requiring larger antennae
can conveniently use the east and west faces of the
satellite for mounting deployed antennae. This
arrangement has the incidental advantage that the 'feed
horns' illuminating the reflectors with radio energy can
be fed by short, low loss waveguide, connections to the
transponders mounted behind the earth facing side.
This arrangement is likely to be favoured for many
direct broadcasting satellites which require relatively
large antennae and low output-loss arrangements. Except
for the very simple, early, communications satellites,
spin-stabilised designs have all had to use antennae
deployed from the de-spun north face.

2.9.5. Attitude Control and Propulsion

Attitude control of the spin-stabilised satellite is
achieved by orientating the spinning axis in the north-
south line and controlling the phase of the de-spinning
rotation so that the antennae point steadily in the
desired direction. Optical detectors are mounted on
the spinning drum which detect the edges of the earth
and this information is used to control the phase of
the de-spinning motor. For very accurate pointing,
directional information is derived from the arrival of
radio waves from the earth and this is used to control

the antenna pointing by actuating mechanisms behind the
reflectors.

In the case of the body-stabilised satellite, attitude
control can be achieved in a number of ways using
internal spinning wheels. In one design, two fixed
momentum wheels are provided set at an angle to each
other. Motors are provided so that each may be speeded
up or slowed down under electronic control. Using the
wheels in combination the spacecraft may be rotated
about any desired axis. Earth sensors are used, as in
the case of the spin-stabilised satellite, to derive
orientation information. Radio frequency sensing and
control of antenna pointing can also easily be achieved.
The momentum wheels supply the 'stiffness' required for
satisfactory platform stability. The body-stabilised
satellite also uses sun sensors to control the
orientation of the solar arrays.

An accuracy of orientation of about 0.1^{o} can be
obtained using the earth sensing optical detectors and
gyroscopic control systems described. This accuracy
cannot, however, be maintained constantly as wheel-
momentum 'dumping' and thruster firing associated with
station keeping manoevres cause additional disturbances.
To overcome the effects of these disturbances and also
to attain a higher pointing accuracy, RF sensing systems
are used. These can achieve accuracies of pointing of
about 0.05^{o} maintained through all phases of the
mission.

Whatever the mode of attitude stabilisation used,
satellites require a propulsion system to provide
adjustment of primary angular momentum. Angular
momentum is conserved within the spacecraft system and

internal wheels are only able to absorb external torques
of an amount set by the limits of their rates of
revolution. Since the satellite is subject to continual
torques from solar pressure, magnetic fields and micro-
meteorite impact it is necessary, periodically, to 'off-
load' the wheels and return the attitude control system
to its central control position.

This angular momentum 'dumping' requires that the
propulsion system activates a pair of thrusters pointing
in opposite directions to provide a turning couple.
Sets of thrusters are mounted on the outside of the
body-stabilised satellite to provide for this requirement and they are fired in the required manner by remote
command. The spin-stabilised satellite has an identical
requirement and this is again fulfilled by sets of
thrusters. However, because the main body of the
satellite is spinning, the thrusters may have to
operate in a pulsed mode. Moreover, this arrangement
allows only two pairs of thrusters to provide main and
reserve capability for any direction; the phase of
the firing pulse determines the direction of resultant
torque. Fuel flow to the thrusters is achieved by
pressurisation in the body-stabilised design and by
'centrifugal force' in the spin-stabilised arrangement.

Similar thrusters to those required to provide for
changes in angular momentum are also used for orbit
control. In this case, of course, they are fired to
give an overall increase or decrease in orbital
velocity for east-west station keeping and a change in
orbit inclination for north-south corrections. The
propulsion system may use one or two liquid propellants.
The present designs of satellite use a mono-propellant
system based on the catalytic decomposition of hydrazine

sometimes with a heated catalyst to increase the energy
of ejection. Later designs are favouring the use of a
bi-propellant; un-symmetrical di-methyl hydrazine and
nitrogen tetroxide are often used. The bi-propellant
systems offer the advantage of higher specific impulses
but bring the disadvantage of double feed requirements.

2.9.6. Telemetry Tracking and Command

A satellite in orbit cannot function effectively if
there is no way of monitoring its performance, knowing
where it is and being able to affect its operation by
remote command. These functions are performed by the
telemetry, tracking and command system (TT&C). Although
there is an increasing trend to provide automatic
spacecraft which can take independent action to remedy
problems or optimise efficiency, manned intervention
will always remain an essential feature, even if only
in an over-riding mode.

The TT&C system is a substantial part of the overall
system cost. Several hundred test points in the
average communications satellite are examined by the
TT&C system and converted to electrical, digital
quantities. The satellite's telemetry transmitter
constantly sends this information to the earth where
it is picked up and processed by the TT&C earthstation.
Two transmission systems are usually provided.

The first, which is also used during the launch phase,
works through an omni-directional antenna on the
satellite and is picked up by large earthstations able
to receive the resulting weak signal. Relatively low
frequencies are used for this telemetry signal, usually
between 130 MHz and 4 GHz. The basic objective of this
transmission system is to get a signal to earth even if

the orientation of the satellite is unknown or varying.
Since the transmitter sends its signals in all
directions, pointing is not required. In orbit, this
mode of transmission is available in case of a mal-
operation of the attitude control system. Clearly, the
one facility that must not be lost in an emergency is
the telemetry & control.During parts of the launch phase
the satellite is not orientated suitably for highly
directional communications and this mode is then the
only one available.

The second mode of transmission uses a carrier close to
the main communication frequency of the transponders
and uses the directional antenna provided for the payload. Because a directional antenna is used, the signal
is much stronger at the TT&C earthstation and modest
sized antennae can be used for reception.

The telemetry transmissions from the satellite are used
as part of the determination of the satellite's position.
The TT&C earthstation uses a carefully calibrated
tracking antenna to resolve the angular direction to
the satellite and this information, together with the
distance to the satellite (which is determined by
measuring the time taken for a ranging signal to travel
out to the satellite through the electronics and back
again) is sufficient to fix the satellite's position.
A more accurate determination of satellite position can
be made if two or more earthstations cooperate to make
measurements and this will be necessary as tighter
tolerances on station keeping are introduced to conserve
orbit space.

Commands are sent to the satellite for many different
purposes. There are instructions only sent once, like

the signal for the apogee boost motor to be fired or for
an antenna to be deployed, routine instructions, like
those used to fire thrusters for manoevres or switching
payload sub-systems off for eclipse periods and urgent
instructions sent to switch to redundant equipment.
        The TT&C earthstation contains the equipment for
generating and checking commands. It is usual for the
spacecraft to repeat the command back to the ground after
its reception, so that it can be checked before execution
is authorised.

The control of satellites is a highly skilled operation.
The utmost familiarity with the spacecraft systems is
required of operators; the ideal arrangement is for the
manufacture also to control the spacecraft during its
lifetime in orbit.

The command system uses frequencies and techniques
similar to those used for the telemetry system. Again
provision has to be made to command the satellite in a
'tumbling' mode. On station the satellite's communication
antennae can be used.

2.9.7. Launch Separation to Operation

Between the end of the launch phase proper and the
commencement of operations a communications satellite
goes through apogee motor firing, drift orbit, deployment
of antennae and solar arrays (if required) and attitude
determination and adjustment. The thrusters of the
propulsion system are also operated to correct for
'dispersion' in the launch and apogee boost operations
and to halt the drift when the wanted orbit location is
reached. A plane change is required to get from the
inclined launch orbit (caused by the launch site not
being on the equator) to the zero inclination geo-

stationary orbit. This can be done at a number of
different points in the launch process.

The spacecraft must retain the direction of the apogee
boost motor during firing. Present designs use spin-
stabilisation and, after separation from the launch
vehicle, the satellite is spun-up by firing its thrusters.
Newer designs, employing liquid propellant apogee motors
use body-stabilisation techniques for this phase. Power
for the spacecraft during this period of its life is
provided by the solar arrays. In the case of the body-
stabilised design, the outer panels of the arrays folded
flat against the north and south faces provide
sufficient electricity.

After apogee motor firing, the spacecraft is orbiting at
nearly geostationary velocity and is said to be in a
'drift' orbit. It is then necessary to adjust the rate
of drift by using the propulsion system's thrusters.
How much adjustment depends on the accuracy of the
launch injection into transfer orbit and the apogee
boost motor's impulse. All these processes are
calculated to provide a nominal result, but allowance
has to be made for inaccuracies.

During these early stages of the mission the spacecraft
does not appear stationary as seen from earth. In
consequence telemetry reception, tracking for orbit
determination and commands to the satellite have to be
channelled through a number of earthstations around the
world able to 'lock-on' to the satellite's transmissions
and follow its motion. Several tracking networks exist
associated with the various launch options.

Once a stable drift orbit is established the spacecraft
can be de-spun (if it is a body-stabilised design),
deployment mechanisms can be released, the spacecraft
can be orientated into its correct attitude, wheels
brought up to speed and in-orbit tests commenced.

During the lifetime of a satellite it is sometimes
necessary to re-position it to a different longitude
orbital location. This is achieved by firing thrusters
in the propulsion system to increase or decrease the
satellite's velocity in orbit. The satellite is then in
a drift orbit again and will continue to change longitude
at a constant rate until the propulsion system is again
used to resotre exact geostationary orbital velocity. The
amount of fuel expended on this manoevre depends on how
much velocity change is chosen and this, in turn,controls
how long the satellite takes to reach its new location.
For example, an orbital change of 10° accomplished in 10
days requires a velocity change of 8.5 m/s to start the
drift and an identical amount to stop it. The amount of
fuel expended on this manoevre would be equivalent to
125 days north-south station keeping which requires
about 50 m/s per annum.

As has been stated many times, the mass of fuel carried
by the satellite is a major part of the payload and, all
being well, it determines the life of the satellite. It
is therefore one of the more important questions to be
resolved before the cost of a satellite and the
viability of the system can be determined.

2.10 PAYLOAD SYSTEM DESIGN

2.10.1 Introduction

The payload system of a typical direct broadcast
satellite, required to amplify the uplink signal,
frequency change it to the broadcast band and re-transmit
it back to earth, is shown in Figure 2.8. The payload
is made up of the antenna(e) and electronic amplifiers,
frequency changers etc. needed for the mission.

The payload is specific to the system in which the
spacecraft is to work being tailored to work at the
exact frequencies and power levels specified. The shape
of the uplink and downlink antenna patterns, in
particular, are unique to the requirement as they are
related to the shape of the country to be served. The
amplifiers, mixers and switches may come from a standard
range of components but the filters have to be 'custom
built'.

The payload is designed so that active equipment is at
least duplicated. Where components have a finite lifetime, for example, hot-cathode components, reliability
analysis techniques are used to determine how many
alternate 'redundant' chains have to be provided to give
adequate availability at the end of life of the satellite.
Passive components are not automatically duplicated, but
in common with all the equipment in a satellite, they are
life tested and 'qualified' as suitable for a spacecraft
by being subjected to environmental tests designed to
simulate all the phases of the mission from launch to end
of life.

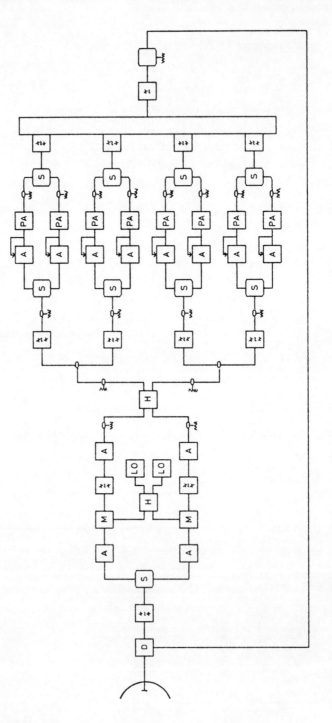

ANTENNA   LOW NOISE   DOWN   WIDEBAND   INPUT   CHANNEL   POWER   OUTPUT
             PRE-AMPLIFIER CONVERTER AMPLIFIER MULTIPLEXER AMPLIFIER AMPLIFIER MULTIPLEXER

FIG. 2.8   TYPICAL COMMUNICATION SATELLITE PAYLOAD

Legend for Fig 2.8

<u>D</u>. Diplexer - a microwave component
            separating signals
            travelling in opposite
            directions in a waveguide

<u>S</u>. Switch - some switches in the payload
           are in coaxial form, some in
           waveguide. Both types are
           mechanical

<u>A</u>. Amplifier - low-noise amplifiers and
             level raising amplifiers
             use the same symbol

<u>M</u>. Mixer - a non-linear component mixing
          together two signals applied
          to it

<u>H</u>. Hybrid - a four port microwave
           component. An input applied
           to any port is split between
           the two opposite ports

<u>PA</u>. Power Amplifier - as explained in
                    the text PAs used
                    for Direct Broadcast
                    satellites use
                    Travelling Wave Tubes

The other component frequently shown is
a circulator. This is a three port
microwave device used to prevent energy
reflected from mis-matches travelling
backwards through the payload. The third
port is terminated in a load resistance
in which reflected energy is dissipated.

Filters are shown in the conventional
manner with two or three 'wavy' lines
to indicate the filter characteristic.
A line through a 'wavy' line indicates
the stop band. All the filters shown
are band-pass except for the harmonic
filter following the output multiplexer
which is a low-pass design.

2.10.2 Low Noise Amplifiers and Mixers
- The Wideband Section

The same antenna is used on the spacecraft for reception
and transmission, and the signal from the earthstation,
at 18 GHz, is separated from the transmit signal by the
diplexer network using directional sensitivity, and
polarisation and frequency discrimination. After
conversion the signal is fed via a two way switch to one
or other of two low noise amplifiers and frequency
converters. This section of the payload amplifies and
frequency changes the whole 500 MHz band of frequencies
received from the ground which may contain up to 5
complete television programmes and associated sound
channels and must be designed to work with a linear
transfer characteristic to avoid inter-mixing of the
separate signals.

In order to obtain adequate quality signals, the first
stage of the low noise amplifier uses Gallium Arsenide
Field Effect Transistors (GaAs FETs). A bandpass filter
is required in front of this amplifier to prevent
unwanted signals entering it. The passband of this
filter is designed only to accept signals in the
required 500 MHz band.

The receiver is a single conversion arrangement, the
18 GHz input signals being downconverted to the 12 GHz
transmit band by mixing with a 5.6 GHz local oscillator.
The local oscillator is derived from a frequency
multiplication chain starting with a crystal oscillator
at about 100 MHz. The crystal oscillator requires to be
frequency stabilised against temperature changes.

The receiver consists of two separate chains for
redundancy. These are arranged so that only one chain
is switched on at a time. A switch determines which
chain is fed with the input signals, but the outputs
are combined in a passive mixing network. This avoids
the additional risk associated with a switch which
always offers the possibility of sticking in the wrong
position. The same arrangement cannot be used at the
receiver input as this would lower the input signals by
a factor of at least 2 and prevent the low noise
amplifier from working effectively.

Two local oscillators are provided for reliability and
are interconnected so that whichever is switched on
supplies both down converter chains. This is achieved
using a microwave hybrid component and the levels out of
the oscillator circuits are designed to accept the losses
involved in such a passive splitter. This arrangement
avoids the possibility that could arise, if each
oscillator supplied one down-converter chain, of a
failure of one receiver and one local oscillator leading
to complete failure.

The power level of each television programme entering
the low noise amplifier is about 1.7×10^{-9} watts and
after amplification and frequency conversion the level
is about 32,000 times greater, at about 5.5×10^{-5} watts.

2.10.3 Channel Filters and Power Amplifiers
- The Channelised Section

The signals are separated before further amplification
by band-pass filters. The filters are each fed with all
the signals received from the uplink earthstation and
each selects only the channel it is tuned to accept.
The arrangement of channel filters fed from the

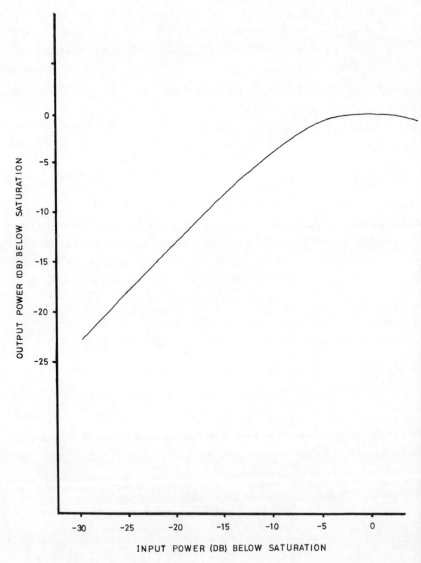

Fig.2.9   TWTA TRANSFER CHARACTERISTIC

composite signal is sometimes achieved by a single
microwave component called a manifold input multiplexer.

The reason for separating out the signals at this point
is because the high power amplifiers available at
present have limited power handling capacity and
because they are non-linear devices. In the case, for
example, of a direct broadcast satellite radiating about
200 watts for each programme, the five channels
allocated would require that a single amplifier handling
them all produce 1000 watts. The space-qualified
travelling wave tube amplifiers (TWTAs) now used are
limited to about 250 watts.

However, the non-linear characteristic of TWTAs close
to saturation is a more fundamental bar to using them
for simultaneous amplification of high power signals.
Figure 2.9 shows the relationship between input and
output signals applied to a TWTA. At low level, the
output signal increases in direct proportion to the
input signal. This is termed the linear portion of the
transfer characteristic. However, as the input is
increased the rate of increase of the output falls and
eventually, at saturation, there is no change in output
level when the input signal is varied. To get the best
efficiency out of a TWTA it has to work close to
saturation but, unfortunately, in this condition several
signals applied together to the input interfere with one
another.

This phenomenon is called intermodulation and occurs
because the amplification of the TWTA varies for any one
signal as it rides up and down on the non-linear curve.
Each signal's position on the curve is varied by the
instantaneous value of all the other signals and
therefore is amplitude modulated by them. This has the

effect of producing 'inter-modulation products' which
rob the wanted signals of power and are transmitted,
causing interference. A mathematical treatment of
inter-modulation product generation is given in an
Appendix.

TWTAs are extremely efficient devices; modern multi-
collector tubes are able, at saturation, to output about
half the applied electrical power as radio frequency
energy. Their drawbacks of limited life, comparative
fragility and need for high voltage supplies are overcome by careful design. Solid-state power amplifiers
are used up to about 4 GHz at present in transponders,
but are limited to powers of tens of watts per module.
Their efficiency is, at best, only 25%. TWTAs are
likely to remain the choice for 12 GHz direct broadcasting satellites for a considerable time.

The consequences of TWTA limitations are seen in the
block diagram of the payload. Firstly, because of the
non-linear characteristic, the working point of the
TWTAs has to be very closely controlled and automatic
level controllers, as shown in the diagram, are used to
stabilise the input levels to the tubes. As the tubes
wear out the input level has to be increased, and so it
is arranged that the stabilised tube input levels can be
altered by command from the ground. The life-time of
TWTAs requires that spare amplifiers are available and
the diagram shows a complete redundant chain for each
channel. Sometimes even more efficient redundancy
arrangements are used which allow any spare TWTA to be
used in place of a failure in any channel.

Following the 'redundancy' switches, the outputs from
the high power amplifiers are combined in the output

multiplexer. This microwave component couples the output from each TWTA through a band-pass filter to a common manifold. For direct broadcasting applications, where maximum output power is very important, low loss output multiplexers are necessary. Even so, for a satellite with five 200 watt channels and typical loss of 18% in each channel (0.7 db) leads to a heating effect of 180 watts in the output multiplexer requiring special thermal arrangements.

A harmonic filter designed to suppress unwanted signals at multiples of the transmitted frequency follows the output multiplexer. A circulator then isolates preceding circuits from the variable impedance of the antenna which receives the radio energy for transmission.

2.10.4 Satellite Antennae

The payload diagram uses the same antenna for transmission and reception of the broadcast programme signals. Sometimes it is necessary to use separate antennae for these two functions but the arrangement leads to extra expense. Even if separate antennae are used, it is necessary to use the transmit antenna as a receiving unit for the beacon signal transmitted by the uplink earthstation often used to control precisely the direction of transmission from the satellite. Separate antennae are required if the uplink and downlink specifications are such that they cannot be met by a single unit.

At any one frequency the size of the reflector of an antenna system is inversely proportional to the narrowest beamwidth required. For example, at 12 GHz a beamwidth of 0.6° requires a reflector aperture of about 3 metres. The size is directly proportional to wavelength for any beamwidth and, therefore, inversely proportional to frequency.

The WARC '77 satellite direct broadcasting Plan calls for elliptical or circular shaped beams from the satellites to illuminate the coverage areas. There are basically two ways of generating elliptical beams: using a single feed horn to illuminate a non-paraboloidal reflector or several feed horns together illuminating a paraboloidal reflector arranged to synthesise an elliptical beam. Both have been used in current broadcast satellite designs. A feed horn is a flared section of waveguide (which is simply a pipe down which radio waves can be made to travel) arranged so that the radio waves can pass from a 'guided' condition to the free space environment with as little loss as possible.

The choice of which antenna design to use has to be made for each application. The single feed approach avoids the need for the branching arrangment needed in the case of multiple feeds, and therefore is lighter and more efficient. However, the design and fabrication of non-paraboloidal reflectors is avoided by the multiple feed method. The extraction of directional information is easy in the case of the multiple feed system as the relative strengths of the beacon signal received from four of the feed horns can be used to indicate mis-pointing. The single feed point arrangement makes use of a device called a 'mode extractor' in the waveguide connection to the antenna to extract similar signals indicating the direction of reception of the beacon signals. High-efficiency single feed antennae usually require dual reflectors, multiple feed antennae do not.

Although the WARC '77 Plan (Chapter 5) does not require
them, 'shaped' and 'weighted' beams are increasingly
of interest for broadcast applications. A 'shaped' beam
is one contoured to follow the coverage area more
exactly than an elliptical or circular beam and is
achieved by a development of the multi-feed arrangement
outlined above. By controlling the quantity of radio
energy directed into each feed point and the relative
timing or phase of each, non-elliptical coverage
patterns can be generated. It is also possible to
accentuate or 'weight' the energy in certain directions
to achieve, for example, extra flux-density in parts of
the coverage area susceptible to heavy rainfall. The
saving of radio energy caused by avoiding the illumination of geographical areas not requiring service has to
be balanced against the lower efficiency of the more
complicated branching pattern of the beam-forming
network.

The scattering of radio energy in unintentional
directions from a transmitting antenna both wastes
energy and causes extra interference. Such scattering
is caused, for instance, by surface imperfections of the
reflector(s) or by spacecraft structures in the path of
the radio beam. The solution to surface imperfections
is clearly tighter specifications. The removal of all
structures from the beam path is very desirable and
off-set fed antennae are favoured. In this arrangement
the feed-horn(s) is not placed at the centre of the
reflector system, like the bulb in a car headlight, but
to one side. The reflector profile is then designed
exactly to compensate for the feed point(s) and produce
the same beam pattern as the centre-fed antenna. Off-
set fed antennae will be universal for satellites
providing direct broadcast service to the WARC '77

specification because of the very low level of scattered
signals allowed under that Plan.

Chapter 3 will discuss the polarisation of radio waves.
Briefly stated, the lateral direction of the electric
and magnetic fields with respect to the direction of
transmission can be controlled. It is a function of
antenna systems to generate and receive radio signals in
specific polarisations. All antennae generate linear
polarised waves initially, and if circular polarisation
is required, interpose a quarter wave delay aperture in
a feed system. Both senses of circulation or both
directions of linear polarisation can be handled
simultaneously by one antenna if it is equipped with the
necessary feed arrangements.

CHAPTER 3

Propagation

3.1 OVERVIEW

The performance of a radio-frequency system is specified
by the ratio of the powers of the wanted signal and the
combined effect of the unwanted interference and 'noise'
energy at the receiver. The greater the wanted signal
relative to the unwanted component, the better the
system.

If there are no obstructions between the transmitter
and the receiver, that is a "free-space" path, then the
theoretical signal-to-noise ratio may be calculated from
a knowledge of:

the power output of the transmitter
the transmitting aerial gain
the "spreading loss" (inverse square of the distance)
the effective aperture of the receiving aerial
the noise temperature of the receiver
the ohmic losses at each stage

However there are many practical factors that will
further attenuate the signal and an allowance must be
made for:

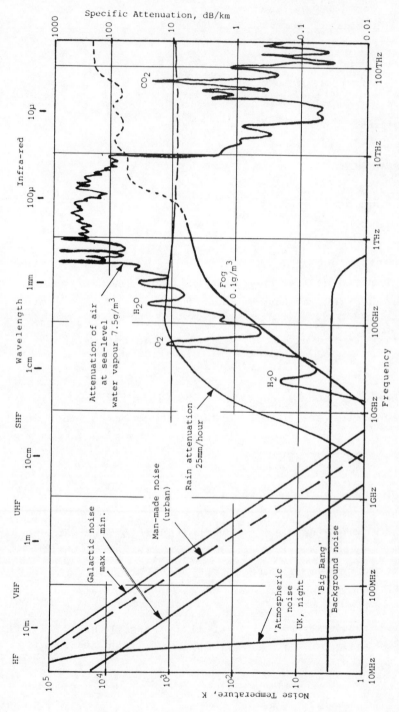

Fig. 3-1. Noise and attenuation from various sources: Noise from CCIR reps. 322, 258, 215; attenuation from CCIR rep. 719.

- buildings, trees and hills
- rain, hail, snow
- fog and cloud
- atmospheric turbulence, inversions
- atmospheric molecular resonance: H_2O, O_2

The noise may come from a number of different sources which include:

- man made noise
- interference from unwanted transmissions
- receiver noise

It should not be forgotten that the effect of this noise depends on the modulation system employed to transmit the picture, sound and data.

Many of these factors are determined by the frequency plan and the technical standards employed. A good plan is the key to the future success of the satellite broadcasting band.

The spectrum of the noise and interference is also important. We shall, of course, be concentrating on the region of the spectrum around 12 GHz but it is interesting to note the trends at other frequencies. Ionospheric refraction, for example, which affects high frequencies (3 to 30 MHz) is not significant at 12 GHz. Similarly, the atmospheric and man-made noises that afflict reception at lower frequencies have dropped to a low level at 12 GHz; even the so-called synchrotron radiation from the hot gases of the Galaxy and radio stars has shrunk below the temperature of the Big-Bang background radiation of about 3 degrees Kelvin (discovered by Penzias and Wilson at 4 GHz in 1965).

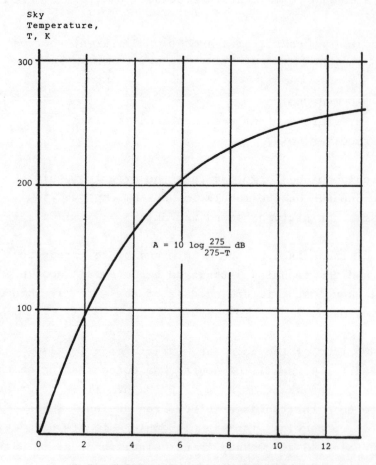

Fig 3.2 Relation of sky temperature and
specific attenuation

More striking perhaps than the fall-off in noise up to
12 GHz is the rise in attenuation at higher frequencies
caused by rain and by atmospheric molecular resonances -
these put an upper limit to the use of the higher
frequency bands that could be used for space-earth links.
Indeed, the heavy absorption by oxygen at 60 GHz renders
the atmosphere almost opaque, so that this frequency
might well be allocated to very short-range terrestrial
use and to communication between space-stations, without
mutual interference between earth and space.

One may conclude that the 12 GHz band is a good choice
for economical low-noise communication between earth and
space, the principal hazard being signal absorption by
rain, given that there is an otherwise clear 'line-of-
sight' path between transmitter and receiver. If there
is any fixed physical obstruction in the path, such as a
tree, a house or a hill, the radio 'shadow' behind the
obstruction will be fairly 'sharp-edged' because the
wavelength is quite small.

3.2 THE EFFECTS OF CLIMATE ON PROPAGATION AT 12 GHz

3.2.1. Attenuation, noise and depolarisation

Given a propagation path that is free of solid obstruction, the signal from a broadcast satellite suffers
little attenuation on its way through the atmosphere
provided that the elevation angle of the satellite is
not very low and that the air is dry. Water, in the
form of invisible vapour, mist, fog, cloud and rain (in
that order) affect the signal to a greater degree
depending on the climatic zone and the local microclimate.

In general, the more water in the atmosphere, the greater the attenuation of the signal but, at 12 GHz, the attenuation caused by mist, fog and cloud, although they may persist for long periods, are usually far less important than that caused by rain, except in rare situations and at very low elevation angles.

Rain has a very important and complex effect on propagation at frequencies of 12 GHz and higher, particularly when the weather is stormy and the rain-drops are not only more numerous but large enough to be compared with the wavelength of the signal and often accompanied by ice and snow. In storms, the size, shape and attitude of raindrops reduce not only the strength of the signal but also the purity of its polarisation. This depolarisation, as it is called, may cause inter-ference from other signals using the same frequency but the opposite polarisation - a practice embodied in the planning for efficient spectrum usage. The effective cross-polar discrimination (XPD) of a receiving aerial may be very high in clear weather but poor in heavy rain (see Section 3.4). In thunderstorms, the change in electric field after a lightning discharge may, by altering the shape of the raindrops, produce a sudden change in XPD. Rain depolarisation is worse for a circularly-polarised signal than a linearly-polarised signal, as will be shown later, but the choice of circular polarisation for satellite broadcasting was made for other reasons; principally for greater freedom in aerial alignment.

Another effect of rainfall on propagation is the increase in sky temperature that accompanies the increase in attenuation, leading to an increase in noise level in the receiver and a further loss of carrier-to-noise ratio. The lower the noise figure of the receiver, the

more sensitive it will be to a rise in sky temperature,
but this effect is not likely to be serious in domestic
reception (Fig.3.2). A thorough study of all these
effects, their severity and their statistical distribution has had to be made before determining the allowances
that must be made in the planning of a broadcast service
that gives adequate signal-to-noise and signal-to-inter-
ference ratios for most of the time. There are three
key elements in the calculation of degradation due to
precipitation:

    (1) Specific attenuation (dB/km)

    (2) Path length through the atmosphere

    (3) The local statistics of precipitation

Given the angle of elevation of the satellite the path
length can be calculated and signal attenuation during
constant rain determined. This simple approach has been
developed further in the Region 2 Conference (1983).
However, this will be inaccurate unless due allowance is
made for the statistics of local rainfall and further
research is needed to establish fade depth and
statistics of system failure. Add to this the similar
problem for the up-link to the satellite and we have a
very large number of factors to consider before establi-
shing a single figure for the required 'noise margin'(or
excess power) and to know how often, and for how long,
that margin will be inadequate.

3.2.2. Methods of measurement

Specific attenuation and cross-polar discrimination may
be found from terrestrial measurements at any desired
frequency and at as many locations as desired, and
correlations may be established with climatic measurements, e.g. rain-gauges. But to find the single-path
results appropriate to a satellite at elevation angles

96

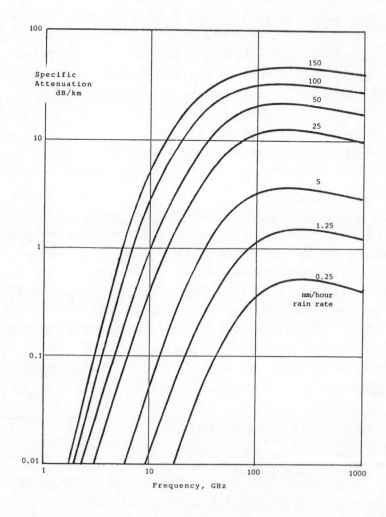

Fig. 3-3. Attenuation by rain

of from zero to 90° requires either a test satellite or
a radiometer. The American ATS-6 and European OTS have
supplied an enormous amount of useful data which is
still being analysed and sifted for planning guidance.
(see Fig.3.2)

3.3 CLIMATIC FADE DEPTH AND DURATION

3.3.1. Rainfall and rain-drop size

The most obvious weather factor that we would expect to
be well correlated with the attenuation of the satellite
signal is the rainfall rate, as measured by a rain-gauge
close to the receiver. We might expect that the
attenuation would be proportional (in dB) to the rain-
rate (the number of drops) and proportional to the
frequency (the drop-size in wavelengths). Both these
ideas give incorrect results for several reasons.

First, the attenuation seems to depend more on the
volume of the raindrops in cubic wavelengths as is shown
in Fig.3-3. Second, the distribution of sizes of rain-
drop depends on the rain-rate itself; in general,
heavier rain contains a greater proportion of larger
drops. This latter relation may need extension when
comparing micro-climates in, say, light rain in the
south-east of England with 'scotch mist' of the same
rain-rate in Scotland, where the drop size may be much
smaller.

3.3.2. Effective path-length

Third, and most important, is the variation of the
'effective path length' which depends on the angle of
elevation, the rain-rate and other factors. Referring
to Fig.3-4, the light rain generally arises from
horizontally-stratified cloud-banks of uniform height,

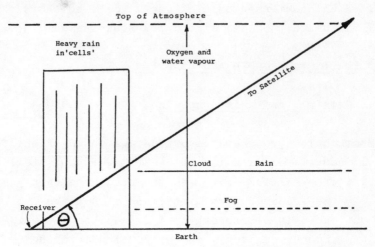

Fig. 3-4. Effect of climate on satellite reception

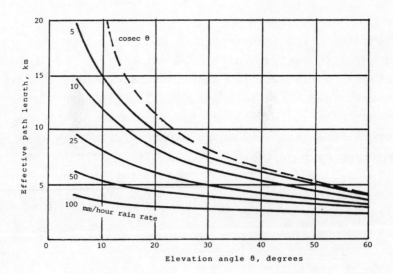

Fig. 3-5. Effective path length through rain
(CCIR Rep. 564)

Artists impression of Olympus 1 spacecraft

(courtesy British Aerospace)

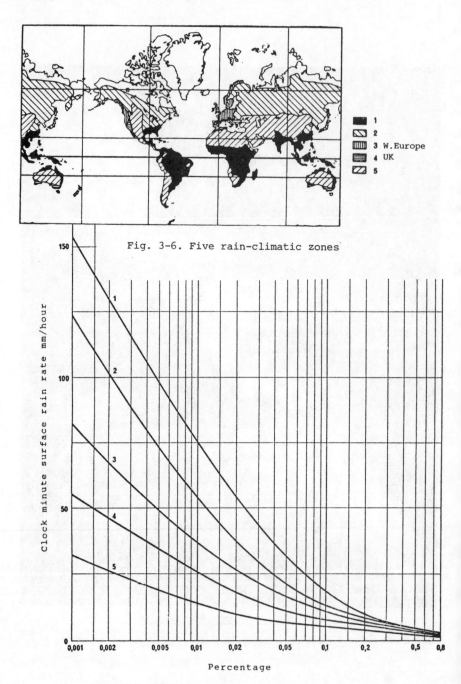

Fig. 3-6. Five rain-climatic zones

Fig. 3-7. Percentage of an average year for which
the rainfall rate is exceeded in each zone

whereas heavier rain arises from taller, discrete "cells" of thunder-clouds, moving past the receiver with a small range of velocities. We can distinguish three separate regimes, giving three different laws relating the attenuation and the angle of elevation. For the uniform layer of precipitation a Cosec law, for the single heavy thunderstorm a truncated Sec law and, for the more usual case of scattered thundery showers, a mixture of the two. Fig.3-5 shows the way in which the effective path length varies depending on the angle of elevation and the rain-rate; these values are based on many measurements by many methods and are suggested by the CCIR as being a good basis for planning services.

3.3.3. Annual mean rainfall

The last two sections show how the actual attenuation may be forecast from a knowledge of the rainfall rate and the angle of elevation of the satellite. But the rainfall rate is difficult to forecast without very detailed spot measurements all over the globe for a long period of time.

As a useful guide, the CCIR has suggested that planning may be based on average statistics for five zones of the globe. These five zones are shown in Fig.3-6, the rainfall distribution for each zone is shown in Fig.3-7 and the consequent attenuation in Fig.3-8.

Although there is little doubt that Fig.3-7 represents a close approximation to the long-term average rainfall, averaged over the zones concerned, a great deal more detail is required about the distribution of rainfall in position and time. The planner may require, for example, the number of times a year that the signal is likely to fade by more than 3 dB for a period of 5 minutes or more

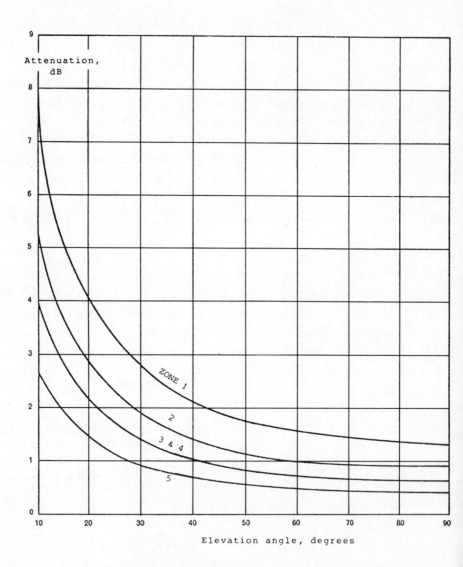

Fig. 3-8. Attenuation assumed for planning satellite
broadcasting at the WARC-77 conference.
The values given are those likely to be
exceeded for not more than 1% of the worst
month (0.25% of the time) in each of the
five rain-climatic zones.

at some particular location among the hills. A full
two-dimensional picture of these probabilities would be
very hard to obtain, but sufficient guidance on each
dimension is available and is described in the next two
sections.

Notwithstanding the complexity of the general calculation it must be stressed that a good broadcast service
may be ensured for nearly all the time over a service
area in a temperate climatic zone by providing a fixed
power margin of only a few decibels to overcome the
effects of bad weather. In more tropical zones, the
heavier and/or more persistent rainfall may require a
larger power margin to provide the same quality of
service. Nevertheless bad weather must be expected to
cause brief interruptions to the service at certain
times and at certain places, particularly at the fringes
of the service areas.

3.3.4. Microclimate

Within each of the five climatic zones shown in Fig.3-6
there is a considerable variation in rainfall from point
to point. For example, the mean annual rainfall within
the small area of the UK varies by a factor of about
five from the near-European South-East to the hilly
North-West. This variation is demonstrated by the
contours shown in Fig.3-9 which deserves careful study.
Figs.3-9(a) and (b) also demonstrate that mean rainfall
and thunderstorm activity have a correlation that is
negative. For example, thunderstorm activity in the UK
occurs mainly in the SE Midlands, whereas the heaviest
mean annual rainfall is gained by the hilly highlands of
Scotland and the Western Isles.

104

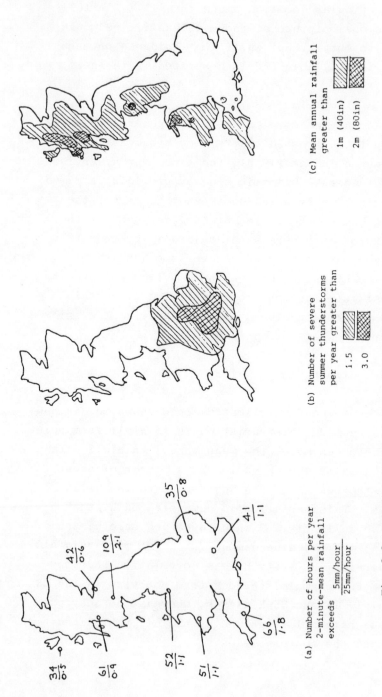

(a) Number of hours per year
    2-minute-mean rainfall
    exceeds $\dfrac{5\text{mm/hour}}{25\text{mm/hour}}$

(b) Number of severe
    summer thunderstorms
    per year greater than

    1.5

    3.0

(c) Mean annual rainfall
    greater than

    1m (40in)

    2m (80in)

Fig. 3-9. Rainfall statistics for the UK from 20 years observations (Slough)

These manifestations of 'microclimate' may be unique, or they may give a guide to the long-term 'randomness of weather'.

The reader is referred to the findings of the Flood Report for a coverage of microclimate in the UK and, by implication, an indication of the importance of micro-climate elsewhere.

Fig.3-9 does not have the detailed contours of the Flood Report but it shows the extremes that can exist in a comparatively small area. In Fig.3-9(a), the striking feature is the comparative uniformity of heavy rainfall rates across the country, justifying exactly the curve for climatic zone 4 (in Fig.3-7). Fig.3-9(b) shows how little is the rainfall accumulated from heavy thunder-storms and Fig.3-9(c) shows, by deduction, that the greater mean rainfall must arise almost wholly from very low rainfall rates. For example, a mean annual rainfall of 40 in.could arise from continuous rain at only 0.1 mm/hr.

3.3.5. Single-point fading statistics

Signal fading, caused by rain-attenuation, depends on a number of factors, as explained in the earlier sections. At any one point the signal fading which occurs there is closely related to the rainfall-rate and the rate at which the weather-pattern passes the receiver. Bearing this in mind, the general characteristics of fading at any one place should be acceptable as a guide to the statistics of fading at any other place (to a reasonable degree of approximation).

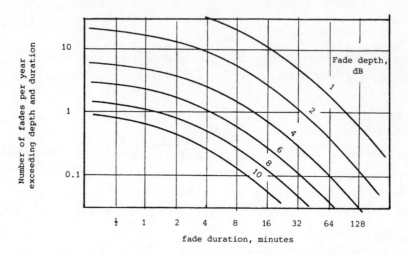

Fig. 3-10. Number of fades per year

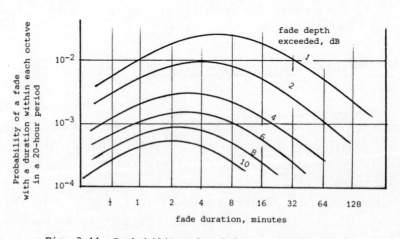

Fig. 3-11. Probability of a fade in a day's reception

These fading statistics were derived from sky-noise radiometer measurements taken at 11.6 GHz and at 29.5° elevation over four years at Slough (Allnutt)

Nevertheless, it is important to find the average short-
term pattern for one particular area, and this has been
done by radiometer measurements at Slough over a period
of 4 years of continuous readings. The full breakdown
of recorded times at or above certain fixed levels is
not given here, but the figures produce the mean annual
number of fades exceeding a given depth and duration as
shown in Fig.3-10. It may be assumed that the same
statistics apply closely to other areas of similar
rainfall density - but this assumption needs verifica-
tion.

Fig.3-11 is designed to show the most likely period of
a fade of a given depth, and how this most-likely-period
depends on the rain-rate. As one might expect, the
heavier the rain, the smaller the rain-cell and therefore
the shorter the period of outage, but clearly it is
possible for a slow-moving thunderstorm to lead to a very
long-period, deep fade on a few occasions. Such isolated
events can be most annoying to a broadcast-reception
viewer and may have to be taken into account in planning
a service with greater weight than the more frequent
fades of lower attenuation.

3.3.6. A worked example for the United Kingdom

Taking very little more than the data already given in
this chapter, it is possible to derive meaningful, if
not completely substantiated, statistics of fading for
different areas of the UK for which detailed fading
figures are not yet available. The first step is to
find the rainfall statistics and the satellite elevation
angle at the receiving site; from these, the effective
path length and the specific attenuation may be found
and hence the attenuation statistics. (With some

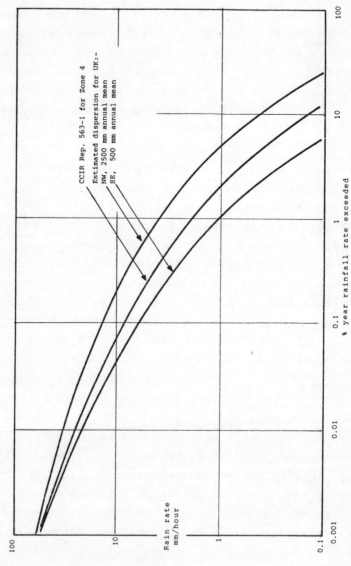

Fig. 3-12. Microclimate: rainfall rate dispersion estimated for the UK

reservation, the same statistics lead to the expected degradation in cross-polar discrimination (XPD) associated with the rainfall - this will be discussed in section 3.4).

The main cause of error in this calculation is the lack of detailed, local rainfall statistics on which to base the calculations; what is needed is more knowledge of the dispersion of climate into a range of micro-climates.

Microclimatic dispersion in the UK can, however, be deduced from the observations shown in Fig.3-9 in the following way. First, Fig.3-9(c) shows how the mean annual rainfall ranges from over 2000 mm in the NW to under 1000 mm in the SE (the greater detail of the Flood Report gives a range of from 2500 mm down to 500 mm). In spite of this large range of 5:1 across the country, Fig.3-9(a) shows that the contribution of heavy rain (over 25 mm/hr) is small and even moderate rain (over 5 mm/hr) cannot account for more than a small fraction of the annual mean - even in the drier parts of the country. The greater part of the rainfall in the wetter regions of the UK must therefore fall at lower rainfall rates, say around 1 mm/hr and below; readers who have experienced the wetting properties of a 'Scotch mist' will perhaps more readily accept this conclusion, which may now be expressed in a more useful form.

The average rainfall statistics for the UK, taken from the CCIR curve 4 of Fig.3-7, have been redrawn on a different scale in Fig.3-12, together with two extra curves showing the extremes of microclimatic dispersion for the SE and NW of the UK. As explained above, the dispersion is greatest at the lower rainfall rates and the curves have been drawn to show this fact and to

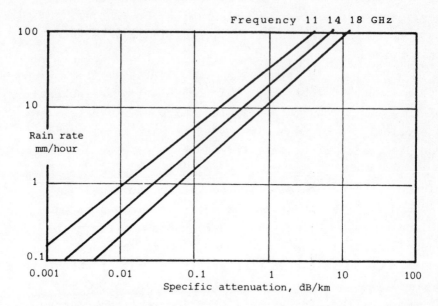

Fig. 3-13. Attenuation at three frequencies

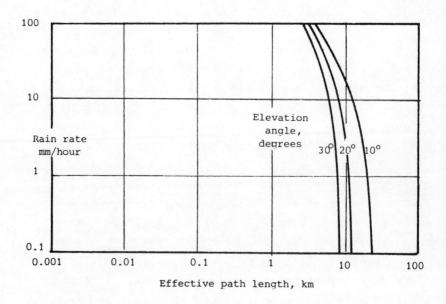

Fig. 3-14. Effective path length through rain

integrate to the appropriate annual rainfall for the two
extreme sub-regions.

Now, given the fading depth and duration figures as
measured at Slough (Fig.3-10) it seems reasonable to make
the assumption that there is a general relationship
between fading depth and duration. In this case we can
make a unifying step forward and associate the rainfall
rate with the fading statistics in other parts of the
country. We can then apply these calculations to the two
extreme sub-regions of the UK and at the three frequencies of possible interest in the satellite broadcasting
band, 11, 14 and 18 GHz.

Short of a more specific survey at these two frequencies
and, over a long period, at these two locations within
one climatic zone, one can interpolate between accredited
values to arrive at a useful working system for planning
purposes. So, re-shaping the graphs of Figs.3-3 and 3-5,
as shown in Figs.3-13 and 3-14, a more specific set of
relations may be drawn up to show the statistics of
fading for the two disparate areas of the UK in the SE
and the NW.

The results of this 'speculative analysis' are shown in
Fig.3-15, the ordinate being the fade-depth and the
abscissae giving three basically-unrelated statistics
that appear to be fairly-well related by the measurements
already discussed.

It may be said that one five-year average is very
different from the next five-year average - so different
that the fine structure of the statistics becomes
meaningless - but the overall trends and variations are
the essence of forecasting, and these must be studied

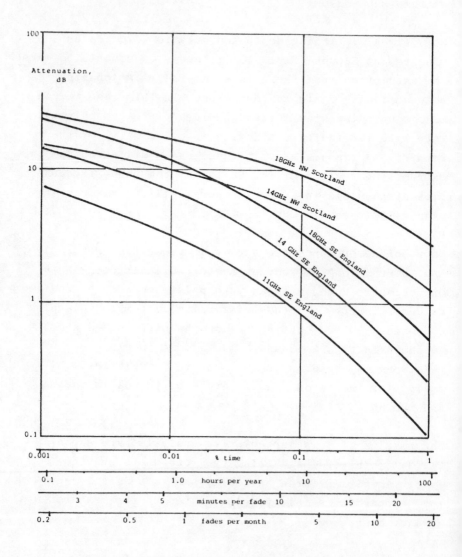

Fig. 3-15. Estimated fade depth and duration in UK in
reception of a satellite at 10°E. CCIR 1978 data
extended and resolved using rainfall and
radiometer observations at Slough

closely, possibly in more detail than they deserve, in
order to discover a general trend of more general use.

We may use Fig.3-15 to make a few general points about
fading margins.   For example, the broadcast up-link,
which must operate in the 17.3-18.1 GHz band, would be
sent more economically from an Earth-station in the SE
than the NW;   a similar analysis could be done for other
areas and other countries.   If, for example, the up-link
were required to maintain a C/N of 25 dB or more for all
but 1 hour a year (say one five-minute fade a month), a
noise margin of 12 dB would be needed in the SE, rising
to 17 dB in the NW.   On the other hand, a mobile up-link,
using the 14 GHz band to access a satellite such as OTS
or ECS, would have an advantage of some 7 dB over the
use of 18 GHz in the NW of Scotland, although it would
not be required to maintain such a high standard as a
fixed up-link.

3.3.7. Satellite power (see also Chapters 4 and 5)

To put the effects of fading into perspective, we may
take a more specific example of the worst that may
happen at a well-placed domestic receiving site, with a
0.9 metre dish, near the centre of England.   Technical
development since 1977 has been such that we may now
assume that the receiver noise figure is several
decibels better than the 8 dB assumed in the planning,
leading to an expected 21 dB carrier-to-noise ratio in
clear air near the centre of the beam.   We also know
that the carrier-to-noise ratio may be allowed to drop
as low as 8 dB for short periods before the picture
becomes intolerable .
So we may consider that a 'drop-out' in service at a
good site with a standard receiver requires a power loss

of at least 13 dB. Interpolating from Fig.15 for a
12 GHz signal near the centre of England, a fade depth
of 13 dB is exceeded for about 0.001% of the time or an
average of about two fades of two minutes duration every
year. The number of such fades that would occur during
normal viewing hours will be correspondingly lower,
perhaps only once every three years of so. Climatic
interruptions of this sort are probably less common and
less disruptive than many other possible causes of
service failure.

The satellite service will also be subject to interference from other transmissions. However, for example,
those who live in the outer parts of Europe will be well
protected from interference from co-channel and adjacent
channel frequencies.

Furthermore the effects of interference have been shown
to be negligible when the F.M. carrier is well above
threshold failure. Interference only becomes a problem
when receiving a weak signal; for example when receiving
a signal in the overspill areas. Thus the WARC plan was
perhaps a little too cautious and, as we shall see later,
the next planning conference (RARC) adopted slightly
different standards.

If the vision signal is scrambled then interference will
be even less of a problem.

Thus improvements in receiver design, geographical
advantage and the cautious approach adopted in 1977 now
prompt reconsideration of the required satellite power,
or alternatively, antenna dish diameter. It would seem
that a dish diameter of 0.6 metre could give good
picture quality and sufficient discrimination against
interference. Furthermore such an antenna would be more

acceptable from an aesthetic point of view and be easier
to install and more likely to maintain its alignment.

However financial issues may dictate that the advantage
is used to reduce satellite power. Unfortunately large
reductions in power cause rapid degradation in the
service. These lower power transmissions would also be
more prone to interference from those not adopting this
reduced standard.

Thus a suitable compromise might be to reduce the power
of a service originally requiring 200 watts to say 100
watts and have receiver dishes of 0.6 metre.

With this signal strength a fade depth of   7 dB would
make the picture unusable but the frequency of this
would depend on the geographical location of the
receiver. From Fig.3-15 we see that this will interrupt
the service for say 3 minutes every three months, that is,
12 minutes per year for the fortunate parts of the U.K.

It should be remembered that terrestrial broadcasting
has set a higher standard of reliability and that concern
over the loss of an important broadcast is not commensur-
ate with the duration of the break.

The debate so far has centred on the likelihood that a
dramatic failure will occur near the centre of the beam
and consequently near to one of the major conurbations.

However it may be desirable to consider the public
service aspects before departing from the WARC specification. What about those who live on the periphery of
the beam where the signal is by definition 3 dB down?
We must also build into the calculation tolerances for
the other parts of the system. Last but not least we

116

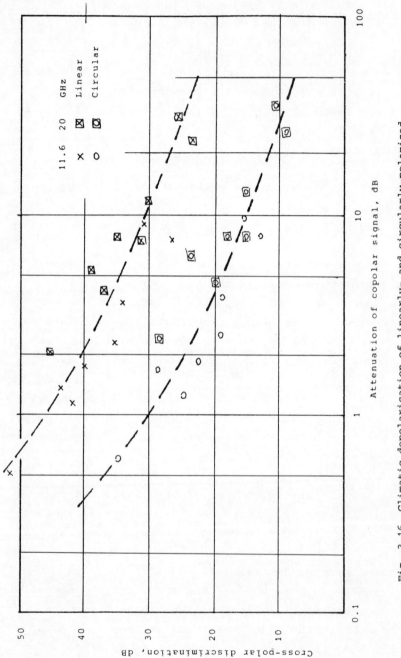

Fig. 3-16. Climatic depolarisation of linearly- and circularly-polarised signals at two frequencies. Measurements taken in four separated areas in Europe and North Africa (ESA)

have to consider what will happen on the typical overcast
day. Will the standard of service match the level of
quality expected from the new technical standards that
will be employed? This aspect of the subject is
considered in greater detail in the next chapter.

3.4 CROSS-POLAR DISCRIMINATION

The effects of precipitation on the polarisation of the
propagating signal have been well investigated and the
correlations with attenuation are fairly well understood.
A review of the statistics of attenuation and XPD for
several climatic zones has been made and reveals a
general relationship between XPD and attenuation
probability as shown in Fig.3-16 for both linear and
circular polarisations.

Most revealing is the difference in performance between
the two alternative forms of polarisation. Of the two,
circular polarisation has been chosen for satellite
broadcasting on account of its greater tolerance of
aerial positioning, even though the alternative of linear
plane-polarisation would have afforded some 10-15 dB
greater discrimination in poor conditions. In spite of
this apparently bad choice, it has been possible, as
shown in the chapters on planning, to accommodate this
degree of cross-polar coupling without undue interference
from co-channel transmissions.

Considering that the data of Fig.3-16 have been collected
from four very different areas in Europe and Africa and
for two very different rain-rates and at two frequencies,
11.6 and 20 GHz, the spread of results about the broken
lines is quite low. It must be said, however, that freak
conditions sometimes occur where the depolarisation
seems uncorrelated with the precipitation.

3.5 INTERFERENCE TO SATELLITE BROADCAST RECEPTION

Although the techniques of frequency re-use are matters
of planning it is useful to include some discussion of
the impact of propagation on planning in this chapter.
The reader may find it necessary to refer to some of the
figures relating to planning matters during the following
discussion.

3.5.1. Frequency re-use in the down-link

As will be explained in the chapter on planning, the
manipulation of three variables - orbital separation of
satellites, frequency separation of channels, polarisation separation of channels - is directed to getting the
best use out of the available channel-space, in spite of
the worst propagation defects, with the least interference. Included in the 'propagation defects' are the
defects of the transmitting and receiving aerials in
their capacity to discriminate between two orthogonal
polarisations of transmissions.

The weakest link in the chain is probably the receiving
aerial and a generous allowance has been made in the
planning to accommodate this assumed defect; but it is
important to remember that depolarisation of both co-
and adjacent-channel transmissions may lead to some
interference in extreme climatic circumstances. For
example, it may arise from the depolarisation of the
side-lobe radiation directed to a distant area from the
same satellite; or it may arise from the near-axial
radiation from an adjacent satellite to an area close to
the receiver, and therefore dependent on the receiving
side-lobes, in which there is little discrimination
between polarisations.

Successful planning requires multiple re-use of channels, close spacing of channels and close spacing of satellites in such a way that interference is just far enough below the acceptable limit from as many sources and for as much of the time as possible.

The ability of the receiving earthstation to discriminate against unwanted signals is mainly determined, as in the uplink case, by the antenna diameter. For a small antenna of 0.9 metres diameter, suitable for direct broadcast applications, the gain factor (and relative isotropic gain in decibels) at 12 GHz is approximately:-

Angle from beam centre (degrees) θ	Gain Factor	Gain dbi
0	7000.0	+ 38.5
6	49.9	+ 17.0
10	13.8	+ 11.4
15	5.0	+ 7.0
20	2.4	+ 3.9
30	0.9	- 0.5
40	0.4	- 3.7
50	0.2	- 6.1
70	0.1	- 9.7
90	0.1	- 12.5

$(G = 52 - 25\log\theta - 10\log$ diameter/wavelength$))$

As this table shows the signal from an adjacent satellite orbital location is discriminated against by a factor of 7000/40.9 = 141 by the receiving earthstation antenna but this is nothing like sufficient to give perfect pictures, which need a ratio of something like 1000 : 1 (although as suggested earlier there is some doubt about this figure in the light of more recent experiments).

To overcome this, the WARC '77 Plan uses the same
frequencies from adjacent orbital locations only to
serve widely separated coverage areas so that the
satellite antennas themselves direct the radio energy
away from the area of potential interference. Opposite
polarisation signals are also employed to give added
separation between wanted and unwanted signals. (Note:
orbital separations are measured as angles at the earth's
centre. Strictly speaking the angle of arrival of
signals from satellites in adjacent direct broadcasting
slots is not exactly 6°. However the difference does
not affect the outcome of this section).

Interfering signals from terrestrial transmitters,
because of their proximity to the satellite earthstation
compared with geo-stationary orbit distances, can be a
serious problem. An Appendix analyses the case of
interference to a direct broadcast system from a terrestrial link and shows, theoretically, that band sharing is
not possible. This result does not hold in practice as
numerous cases exist, particularly in the USA, of earthstations receiving signals from satellites in shared
frequency bands. It is found that local screening of
earthstations can, with careful planning, protect them
from terrestrial interference.

3.5.2. Frequency re-use in the up-link

The up-link, or feeder-link, to broadcast satellites
in Region I is likely to use the frequency band 17.3 -
18.1 GHz; this band is the same width as the down-link,
800 MHz, and may therefore carry an arrangement of
channel assignments exactly the same as the down-link but
transposed up by 5.6 GHz. At one time, however, it was

thought that, because of the greater performance
available at the up-link ground station, giving greater
discrimination between adjacent satellites, the up-link
could therefore be squeezed into a 400 MHz band by
further frequency re-use. Although such bandwidth
reduction was technically possible, it raised so many
extremely difficult problems that a full 800 MHz up-link
band has been sought. The most likely allocation for
this purpose is the band 17.3 - 18.1 GHz.

Unlike the down-link band, however, the up-link band is
not dedicated primarily to broadcast feeder-links -
there must be sharing with other feeder links and certain
rules must be observed.

Apart from the problems of sharing with other services,
the interference allowable between broadcast up-links
has different constraints from that of the down-link,
even though the frequency-channelling may be identical.
The principal constraint is that the performance of the
up-link, both in signal-to-noise and signal-to-inter-
ference ratios, should be about ten times better than the
limit for the down-link. Considering first the noise
performance, there is less benefit to be gained by low-
noise receivers in satellites than those in domestic
installations, because the latter is looking at a cool
sky whereas the former is looking at a transmitter on a
warm Earth. It is usual to consider an up-link noise
figure of the order of 5 dB, giving a noise temperature
of about 1200 K (900 for the receiver and 300 for the
Earth).

Turning to interference, the greater directivity and
stability of the up-link transmitting aerial will enable
each country to pick out its own satellite position with
great accuracy and not spread any interfering signals

towards any adjacent satellite. Now each satellite position may contain several other satellites. Some of these will be receiving on the same frequency and some on the adjacent channel to the channel we wish to watch. There are two defences against interference to our up-link transmissions. First is the discrimination given by the directivity of the satellite receiving aerial and second is the cross-polar discrimination given by both the satellite and the earth-station aerials. Both these defences have to contend with the attack of depolarisation from the climatic conditions over our up-link ground-station. This is because, however good the equipment, heavy rain over the up-link earth-station of one country will depolarise the signal reaching its satellite position and, by the arrangement of the planning, will give direct interference into the up-link receiver on the adjacent channel directed towards an adjacent country.

While the down-link receiving aerial can provide the requisite 15 dB or so to minimise this level of interference, the up-link must find a protection of some 25 dB in spite of a level of atmospheric depolarisation that gives an XPD of lower than 25 dB for about 0.1% of the time.

3.5.3. Up-link planning considerations

Many studies have been performed on the interactive effects of mutual interference between large numbers of up-link systems. The general conclusion is that satellites can be worked at the minimum orbital separation if satellites of the same type are grouped together. The separation of satellites varies from 3° to 6° depending on the types of service and frequencies. The WARC '77 Plan for direct broadcast satellites uses 6°

separation. It is intended that further World or Regional Administrative Radio Conferences will be held to work out Plans for direct broadcast up-links. Up-links for other systems are negotiated case by case.

Interference from up-links is caused by radiation from the earth-station in directions other than towards the satellite. Standards are laid down to limit radiation in the 'wrong' direction: currently the International Telecommunications Union (ITU) limits the gain factor with respect to the angle from the centre of the beam of a typical 10 metre diameter antenna used for an up-link at 18 GHz (gain over isotropic radiator in decibels is also given:-

Angle from beam centre (degrees) θ	Gain Factor	Gain dbi
0	2000000.0	+ 63.0
6	18.0	+ 12.5
10	5.0	+ 7.0
15	1.8	+ 2.6
20	0.9	− 0.5
30	0.3	− 4.9
40	0.2	− 3.1
50	0.1	− 10.5
70	0.04	− 14.1
90	0.02	− 16.9

$(G = 32 - 25\log \theta)$ for $\theta > 1^{\circ}$

Although the gain factor falls very rapidly with angle, an earthstation transmits a powerful signal and interference into terrestrial microwave links can be severe. By way of an example an Appendix shown that the signal from a terrestrial transmitter using, say, half a watt

of power and a 1 metre antenna would be liable to interference from an earthstation working to a satellite at
30° elevation, in line with the terrestrial link and the
same distance away from the receiver as the wanted
transmitter.

3.5.4. Out-of-band interference
Harmonics at R.F.

Certain frequency bands are allocated for the use of
industrial, scientific and medical (ISM) processes
involving high flux-densities of radiation, mainly for
localised heating purposes, for example, fast setting of
modern adhesives in the furniture industry. The most
significant of these bands for satellite broadcasting is
the 2.45 - 2.5 GHz band and two fairly recent additions
to the users of this band are microwave ovens and
proposed solar power satellites.* By the nature of the
microwave power generators used in the ISM bands,
harmonics are plentiful and, though regulations exist
for their limitation, they may give rise to interference.
For example, the 5th harmonic of 2.455 GHz lies close to
Channel 30 at 12.2837 GHz. Although the regulations and
limits may be sufficient in nearly all cases, there will
be the exception where, for example, the receiving aerial
is looking past a high-rise block with microwave ovens
in the canteen on the top floor. If solar power satellites, beaming high powers to earth at 2.455 GHz, become
a fact, there will be some harmonic radiation from the
power receivers that may need careful control to avoid
interference.   * Hence the name 'Radio Gastronomy'

3.5.5. Image and I.F. interference

As described in Chapter 5 , the choice of intermediate
frequency for the domestic receiver is dictated mainly

by considerations of tuning range, stability and noise performance; these narrow the choice to frequencies around 1 GHz. The local-oscillator frequency is similarly chosen on the lower side of the signal for better stability. Originally, the planning assumed that a 400 MHz band would suffice, using one or other of two alternative frequencies, depending on whether the country's assignments were in the upper or the lower half of the 800 MHz-wide transmission band, and leading to two possible 'image' frequency bands. More recently, however is has been seen to be quite practicable to use an i.f. band 800 MHz wide from, say, 950 - 1750 MHz, so that the receiver would be tunable over the whole band with a single local-oscillator frequency (see Ch.5).

Image-frequency rejection may be obtained only by the attenuation of a signal-frequency filter; this may be designed to give a high rejection by using a waveguide filter 'cutting off' at the image frequency. The rejection that can be obtained in this way is limited to a value of about 80 dB over the whole band but, assuming some directivity from the aerial at the image frequency, the level of radar transmissions at the image frequencies should be insignificant in all except the most severe cases e.g. very large radar transmitters or airborne radar actually passing through the sight-line of the aerial (see Appendix).

Interference in the intermediate frequency band may arise from ground-based transmissions such as DME and TACAN at both civil and military airfields; these devices, designed to assist aircraft in finding direction and range from the airstrip, will radiate high power pulses at a low angle. Although this radiation is unlikely to be passed by the r.f. side of the receiver front-end, the down-lead carrying the i.f. output of the mixer from the aerial assembly to the tuner inside the house may pick up some of these pulses. Such interference may be reduced to negligible levels by ensuring that the gain in the front-end unit is great enough to 'swamp' the pick-up on the down-lead.

3.5.6. Interference from satellite broadcasts and receivers

When the WARC-SB Plan is fully implemented, each country will have millions of domestic aerials, all pointing towards the same satellite position and carrying small oscillators at about the same frequency. If the rejection of the r.f. and mixer circuits to the back-radiation is not more than about 30 dB this beam of narrow-band 'noise' directed skywards needs to be taken into consideration.

In the reverse direction, the harmonic radiation from the down-link transmitters in the satellites may be a source of interference to the radio-astronomer. One of the continuing studies of radio-astronomy is the measurement of very low levels of radiation from certain specific molecules in clouds of gas in interstellar space. One such molecular radiation comes from Ammonia, which resonates around 24 GHz, close to the second harmonic of Channel 15 near 12 GHz. Bearing in mind that radio-telescopes have a high directivity, their receivers need to integrate the signal over long periods of up to an hour while 'tracking' a radio-star across the sky, the chances of interference from a satellite are low, but if the worst case is calculated, the required harmonic attenuation at the satellite's aerial filter is over 100 dB. A more detailed description of the problem is given in the Appendix.

3.6 OTHER OBSTRUCTIONS TO PROPAGATION AND RECEPTION

3.6.1. Shadowing by hills, buildings and trees

The angle of elevation of the satellite serving a
particular point may be found from the graph in Chapter 2
or from a simple computer calculation. The highest
elevation possible is somewhat less than the co-latitude
of the receiver (depending on the westward offset of the
satellite from the longitude of the country served) in
order to avoid the worst effects of the equinoctial
eclipses as explained in Chapter 2. The UK satellite at
31 degrees W longitude has a bearing of about 35 degrees
west of south and about 24 degrees elevation above the
horizon as seen from London. Over the whole of the UK
the elevation varies from 28 degrees in Scilly to 17
degrees in Shetland.

It is natural to imagine that, at these angles, many
objects will be in the way. It is difficult to find out
in detail just how many domestic installations will be
'in shadow' without what the planners call a 'site-test'
using a low-power transmitter to find the 'coverage'.
But such a site-test is clearly almost as costly as the
real thing, so some other method must be sought.

Referring to the Appendix it can be seen that, near the
equinoxes when the Earth shadows the Sun from the geo-
stationary orbit at night, the Sun passes behind the
geostationary orbit by day. So it is that, on March 2nd
at about 2.30pm and at 2.00pm on October 13th the Sun
lies behind the position of the UK satellite for a point
near London. (see Fig.5.7).

Using this knowledge, a survey was carried out in
October 1978 to find how much shadowing was a problem

at the homes of some 450 individuals in the broadcasting industry in the UK. The results could not have been forecast, even by a close inspection of a very large-scale map, because trees and buildings are not shown in height.

The survey revealed many interesting features and covered a wide variety of domestic situations all over the UK. The most important result was that, within the limitations of a small sample, the proportion of homes in the UK that were completely shadowed from the UK satellite was less than 0.5% from all causes. The principal shadowing in towns was from buildings, either from shadows cast by the building on to a dwelling or by the fact that the dweller lived in the building and faced north. In either case, the solution to their problems would be a localised distribution system with a communal aerial on the offending building. Secondly, trees were the offenders in only a very few locations. (Trees may be protected and beyond the control of the shadowed homes). Lastly, hills were not responsible for any shadowing; although it was thought that many valley-dwellers would be out of sight of the satellite, it was surprising that all of them in the survey lived sufficiently far up the sunny side of their valleys to stay in the sun at least on October 13th.

The survey revealed some anxiety on the part of the audience and local authorities towards the prospect of large numbers of 3-foot dishes on their roofs or in their gardens; both from the aesthetic and the windage points of view. Local authorities may organise small distribution systems to avoid the environmental pollution of individual aerials. Of course, the roof is not the preferred site for the receiving aerial unless all else is in shadow; many favoured the garage wall or the ground itself. Indeed, a potential viewer in Shetland said that a 3-foot dish on the house would be unnecessary and dangerous as the wind velocity can reach 160 mph and the remaining trees are fairly short. As has been mentioned in Chapter 5 , efforts will clearly be made to design the domestic aerial with these points in mind, so that their propensity for catching the eye and the wind may be minimised. A great improvement on the current design of aerials is possible.

3.6.2. Obstruction by aircraft

An interesting observation was made in the survey regarding aircraft shadows, which had previously been dismissed as negligibly small and sporadic in nature. This may be true in general but not so if we consider those living along the shadow of the glide-paths into a large airport such as London Heathrow, where there are two parallel paths extending over some ten miles of urban dwellings, over which aircraft descend regularly, slowly and very precisely before landing. The result may be that, for the unfortunately placed, there will be a series of 'drop-outs' at half-minute intervals for much of the day, each drop-out lasting, perhaps, half a second, with a short buzz or flutter on either side. The degree to which these drop-outs may be made acceptable is clearly more a matter of receiver design than arranging a cable-feed system. It is thought that, providing the receiver is designed to maintain synchronisation during the drop-out, a suitable 'muting' circuit will remove the worst of the effects of aircraft shadowing.

Chapter 3

References

No.	Subject	Book reference
1	Results of tests and experiments with the European Orbital Test Satellite. Conference, London, 8-10 April 1981, I.E.E. Conference Publication No.199, London, 1981	3.2.2
2	Flood Studies Report, Vol.5 (Maps) Natural Environmental Research Council, London, 1975	3.3.4
3	Further contributions on precipitation attenuation were made to the CCIR in 1982	
4	ITU Final Acts RARC 1983 Part I (precipitation/attenuation formulae)	

CHAPTER 4

The International Plan

4.1 INTRODUCTION

The interests of each nation can only be protected if
DBS is subject to an international plan. As described
in the chapter dealing with the legal aspects the ITU
(International Telecommunications Union) provides a forum
for the international debate which is the essential
prelude to agreement.

The foundations of the plan must be a sound technical
understanding of the problem. Satellite broadcasting
in the 12 GHz band has available a relatively large
bandwidth but unfortunately only low transmitter powers
are practical (although there are those who would
qualify this statement). Fortunately it is possible to
exchange one for the other if Frequency Modulation (F.M.)
is used. In technical terms DBS broadcasting can
benefit from the so-called "FM improvement" (see
Chapter 5) and, using F.M. it was possible for the
planners to save transmitter power without sacrificing
the quality of the received picture.

The general concept was to generate the new television
services on the terrestrial broadcasting television
standards. It might then be possible to adapt existing

receivers using a simple FM to AM converter.

Some farsighted people thought that if new standards were adopted it might be possible to improve the resolution of the picture and provide new services that would make the DBS service doubly attractive. However it was only possible to make vague provision for these developments at the planning stage in the mid 1970's.

It is now necessary to describe these plans in some detail. We will return to the question of sound and television technical standards at the end of the chapter when we will consider what power should, in practice, be used.

This chapter is only intended to give a broad understanding of DBS plans. If further details are required reference should be made to the CCIR official documents of the relevant conferences.

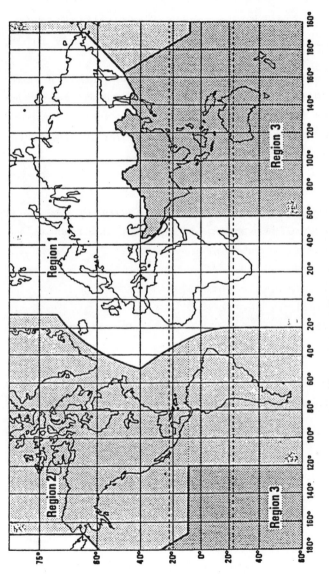

Fig. 4.1 ITU REGIONS

4.1.1 The World Administrative Planning Conference (W.A.R.C.) - Organization

For frequency allocation purposes the International Telecommunications Union (I.T.U.) divides the world into three Regions:

Region 1   Europe, Africa and the Soviet Union
Region 2   The Americas
Region 3   Australia, Asia including China and Japan

In 1971 the Union allocated frequencies in the 12 GHz band for satellite broadcasting but, unfortunately, the frequencies in each Region were slightly different.

Region 1 11.7 - 12.5 GHz: shared in an equal basis with terrestial broadcasting services and the fixed and mobile services

Region 2 11.7 - 12.2 GHz: shared on an equal basis with terrestial broadcating, fixed and mobile services and fixed satellite services.

Region 3 11.7 - 12.2 GHz: as above but without the fixed satellite services.

The requirement that the frequencies are to be shared on an equal basis posed severe technical problems and the need to resolve this quickly was a strong motivation to make a detailed plan for the use of the frequencies without delay, in spite of the fact that satellite broadcasting appeared to be practical only in the distant future.

Accordingly a planning Conference was called in 1977.

In 1976 there was a technical meeting to prepare the
basis for the conference to be held next year and in
1977 delegates from 111 countries arrived seeking
allocations in this band.

The method to be employed was similar to that used at
previous conferences. Politically the only solution
likely to be acceptable was to start with a "fair share
for all" proposal and to modify this to satisfy the
needs of a few countries who could make out a special
case. The proposal was devised by a computer fed with
all the essential technical data. As described in detail
later in this chapter the solution it offered, for
Region 1, was five channels per nation. In Region 3 the
proposal was to provide four channels (the available
bandwidth being less).

Region 2 declined to participate on the grounds that it
was premature. Their planning conference was postponed
until 1982/83.

The International Telecommunications Union (ITU) defined
two areas in broadcasting satellite systems. First
there is the "Coverage Area" where there is sufficient
signal strength to permit the reception of good quality
pictures. Within this area is the "Service Area", where
the interference from other signals is limited to an
acceptably low level. The limit of the Service Area is
usually the geographical boundary of the country using
the broadcasting satellite; the Coverage Area is assumed
to be the area within the half-power beamwidth of radio
frequency signal from the satellite.

4.2 THE BROAD BASIS OF THE REGION 1 AND REGION 3 PLAN

The first step towards a plan was to agree details of the receiver technology which would be employed. Given this it was possible to calculate the signal strength necessary to achieve the required picture quality. It is then possible to establish the satellite transponder power and the antenna characteristics that are necessary to serve the required coverage area.

When devising a plan it is necessary to establish the amount of interference (both co-channel and adjacent channel) that can be tolerated within the service area. DBS allocations must then be made in such a way that the various methods of discrimination between the wanted and unwanted signal combine to give the required protection.

As we shall see later there are three ways in which this discrimination can be achieved. An interfering signal will be reduced by the off-axis characteristic of the transmitting antenna (see Fig.4.2). It will also be reduced by the receiving antenna directivity and by this antenna's ability to discriminate between the two forms of polarisation which may be employed (Fig.4.3).

4.2.1. Satellite Antenna and Coverage Area

The factors that control the strength of a transmission from a satellite at the earth's surface are the power and radiation pattern of the satellite, the distance from the satellite to the receiving point and the amount of the signal which is absorbed in passing through the earth's atmosphere.

The radio emission is a maximum on the antenna axis and falls away to half-power at the angle usually known as the half-power, or 3db, beamwidth ($\emptyset_o$). It goes on falling with angle, normally reaching a first minimum at twice the half-power beamwidth, and then rising and falling cyclically through the 'side lobe' region. The templet defined by the WARC '77 controls the maximum side lobe level which is allowed (Fig.4.2).

The accuracy with which the antenna can be pointed at the earth also has to be taken into account. Particularly at the edge of the coverage area, variations in beam pointing by even 0.05 degrees can reduce (or increase) the signal by a measurable amount. In the case of the UK the edge of the service area moves by about 35 km (22 miles) when pointing errors are taken into account. When estimating coverage area a pessimistic assumption is made about the pointing direction: when estimating the interference potential of a satellite transmission the maximum strength of miss-pointed signal is used as the basis for planning.

If the cross section of the beam is circular then the half-power beamwidth and the boresight completely defines the pattern. However, an elliptical section beam is usually required to achieve an efficient coverage of a geographical area without wasteful over-spill. In this case the orientation of the beam has to be defined. This is measured at the satellite relative to the earth's equatorial plane and a mathematical relation can be derived between this angle and the orientation of the ellipse projected on the earth's surface relative to the line of latitude passing east-west through the boresight.

Working from the satellite antenna pattern, the strength
of the transmission towards any point of the earth's
surface can be calculated. The path distance can be
calculated also from geometrical equations as outlined
in Chapter 2. The loss in the earth's atmosphere can
also be allowed for on a statistical basis, taking into
account the angle of elevation of arrival of the radio
signals at the earth's surface which determines the
length of the path in the atmosphere traversed by the
signals. In the last chapter (Chapter 3) the
statistical basis of propagation through the earth's
atmosphere was considered in detail.

139

Regions 1 and 3

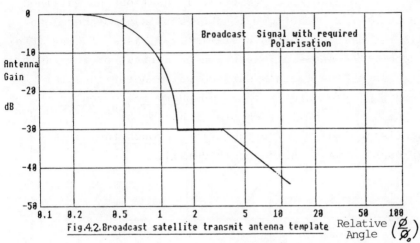

Fig.4.2.Broadcast satellite transmit antenna template

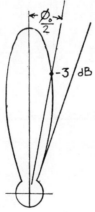

140

Regions 1 and 3

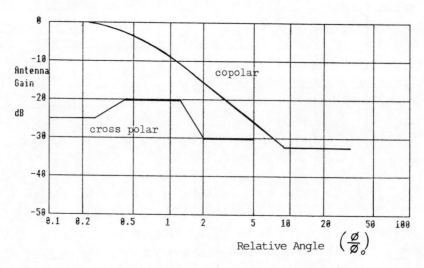

Fig. 4.3 Receiving Antenna Reference patterns

4.2.2. The Receiving Antenna and the Service Area

The receiving antenna was assumed to be a parabolic antenna, or dish, of diameter 0.9 metres. The receiver was assumed to have an 8 dB Noise Figure and a G/T (Figure of Merit) of 6 dB. The antenna performance was assumed to fall within the template given in Fig.4.3.

When all the assignments are taken up the service will be interference limited and the receiving antenna performance plays a major part ensuring the freedom from interference.

It was decided that circular polarisation was the most appropriate because of the simplification in aligning the domestic antenna. Left hand and right hand polarisation could be employed and the receiver antenna discrimination between these two forms of polarisation would be a welcome advantage.

Allowance for terrestrial generated interference follows similar lines to those just described for satellite signals but, in general, gives a much less reliable result; erring on the side of predicting too much interference. This is because the intelligent use of ground cover when choosing a site for the antenna can often minimise this interference.

(Note that the service was to be defined in terms of interference to others, that is the signal outside the service area should not exceed the limiting value. With conventional antenna design the maximum signal strength at the centre of the beam would be about 3 dB higher than the periphery. The signal outside the service area depends on the side lobe performance of the transmitting antenna. Modern technology would seem to be able to

achieve a rather better performance than that assumed
by the conference (see Fig.4.2). This makes it
theoretically possible to reduce the overspill signal.
The practical design will, however, be a compromise much
influenced by cost. The level of each unwanted signal
can then by calculated and added together to form a
composite "interference level".

Sufficient information now exists to check that any plan
of assignments meets the specified interference criteria
within the service areas.

4.3 TECHNICAL DETAILS OF THE WARC PLAN

4.3.1. Antenna

The transmit and receive antenna characteristics were
assumed to fall within the templates given in Figs.4.2
and 4.3.

The plan assumes that the limiting dimension of a
satellite beam is 0.6 degrees.

4.3.2. Reception Angle

In as much as this effects the planning, an angle of
elevation of 20° was assumed.

4.3.3. Satellite Station Keeping

It was assumed that satellites would keep station better
than $\pm 0.1^{\circ}$ in both N.S. and E.W. directions. Similarly
the satellite broadcast beam pointing accuracy would be
$\pm 0.1^{\circ}$. The axis of an elliptical beam would not change
by more than $\pm 2^{\circ}$.

4.3.4. Modulation

As described Frequency Modulation of a PAL or SECAM
signal was assumed with a deviation of 13.5 MHz per volt.

4.3.5 Protection Ratio

The ratio of the wanted and unwanted or interfering
signal, expressed in logarithmic terms, was to be within
the following limits:

- 31 dB for co-channel signals
- 15 dB for adjacent channel signals

Subsequent experience suggests these limits are a little
severe (even more so if the new standards are used and
the signal is scrambled).

4.3.6. Energy Dispersal

To avoid peaks of energy at sensitive frequencies energy
dispersal will be employed. A triangular waveform is
added to the signal of an amplitude such that it
produces a peak to peak deviation of 600 KHz. This
reduces the power within any 4 KHz sample of the
spectrum by about 22 dB and so reduces possible interference. This signal is removed in the receiver (see
appendix for further details).

4.3.7. Propagation

The calculations were chosen on the basis of providing the specified service at all times except for 1% of the worst month in the year

4.3.8. Guard Bands

Guard bands at the edge of the broadcasting band were agreed (11 and 14 MHz).

The presence of these guard bands necessitated the channel bandwidth being reduced from 20 MHz to 19.18 MHz.

4.3.9. Channel Grouping

All frequencies radiated by any one satellite antenna would be within 400 MHz (half the broadcast band) to assist receiver design.

4.3.10. Channel Spacing

It was thought that problems in the combining circuits within the satellite required that adjacent assignments must be greater then 40 MHz apart. A practical solution was to space the allocations by four channels.

4.3.11. Received Signal Quality

It was thought that the transmitter power and hence the signal to noise ratio would, in practice, be limited by the amount of time that the signal is lost because of attenuation when it rains. Clearly too low a power would cause frequent service interruption due to rain and too high a power would be wasteful.

The design target was to achieve <u>14 dB carrier to noise</u> <u>C/N) for 99% of the worst month</u>. The allocated power was to be such as to achieve this carrier to noise ratio when there was 2 dB attenuation due to precipitation,etc.

Using a receiver with a G/T of 6 dB it was calculated that the signal-to-noise ratio in the received picture signal would be about 33 dB (unweighted) for 99% of the time. This would be about 44 dB (weighted) for System I which is a "good" to "excellent" picture quality (see Chapter 5).

It was suggested that the signal should be contained within a 27 MHz channel. Because there was only a small amount of energy at the edge of the channel they could be overlapped by spacing them at 20 MHz intervals without causing visible interference on the picture.

Experiment has shown that a relatively large deviation is possible for the PAL signal (13.5 KHz) and a similar deviation with the C-MAC system proved to be acceptable.

In Europe the number of channels of this type that could be allocated to each nation was limited by the interference. The interference situation will be particularly severe in central Europe if all the allocations are taken up.

4.4 <u>CONCLUSIONS</u>

<u>As a result of these various factors it was agreed that</u> <u>the plan should provide a flux density of -103dB W/sq.M</u> <u>at the edge of the coverage area</u>. Tight tolerances were imposed and an increase in flux density outside this

area (by more than 0.25 dB) would have to be a matter
for negotiation with all concerned.

With a few minor exceptions the plan will ensure
freedom from interference greater than the prescribed
level throughout the service area.

Thus in Region 1 the 800 MHz broadcasting band would
support 40 channels spaced at 19.18 MHz. This can be
effectively doubled if the two forms of polarisation
are used.

Furthermore the receive antenna discrimination was
sufficient to permit satellites to be stationed at
6 degree intervals around the geostationary orbit. The
size of the region is such that there are many time
zones and a number of orbital positions are required.
Most countries sought a westerly orbit because this puts
the eclipse period early in the morning and allocations
for most countries were possible in orbital positions
from 5 degrees East to 37 degrees West.

(Some nations claimed sovereign rights over these
orbital positions but these claims seem to have been
abandoned. In any case most of the Region 1 orbital
positions are over the Atlantic Ocean).

It was decided that it would be possible to assign to
each administration five satellite broadcasting channels.
The footprint of such a service being sufficient to cover
its own territory and not cause interference to its
neighbours.

However, it was recognised that the signal will spill
over on to adjacent countries and such a spillover will
often be welcome (see the chapter covering legal issues).
In which case it would be wise to group the channels on
a satellite so that it would be possible to receive
signals from adjacent countries without repositioning
the domestic aerial.

Furthermore the frequencies assigned to any such country
would ideally be contained within one half of the band
(it being assumed that a practical receiver would not be
built to operate over the full band).

At this point the political pressures for other compro-
mises were applied and the ideal plan described above was
modified in a number of ways to meet the specific requirements of many countries.

4.4.1. Exceptions

There were a number of requirements for "super beams".
There was a proposal from Austria,Switzerland and Federal
Republic of Germany. Similarly Denmark, Finland, Norway
and Sweden each requested two Scandinavian super beams.
Luxembourg sought to cover neighbouring countries;
Andorra to serve France and Spain; Monaco to cover
France; Ireland, UK and Tunisia to cover Algeria, Libya,
Morocco and Mauritania. There was also a bid for a large
beam to provide an Islamic programme.

An attempt was made to provide for all of these requirements but this caused excessive interference and ,for
example ,the co-channel interference fell to 10 dB below
the specified limits.

Administrations were asked to reduce their requirements.

In the end, after much discussion, most of these requests
were withdrawn. Exceptions were made for the Scandinavia
block, Tunisia and, to a limited extent, the Vatican City
and the Islamic requirement. Yugoslavia had ten
programmes to deal with their language problems. A
number of other variations were made in order that an
acceptable compromise could be made and ultimately all
requirements were met with interference in all cases
within 3 dB of the specified limits.

The broad pattern of the plan within Europe is
illustrated by the diagram in Fig.4.4 and it should be
noted that the interference situation prohibited the use
of all available frequencies.

A complete table of assignments, in alphabetical order,
follows. For many purposes it is useful to see who
share orbital positions and a table in this form is
included in the appendices.

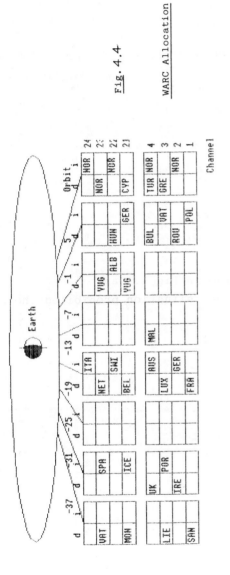

Fig. 4.4

WARC Allocation

The diagram shows the allocation for a selection of countries in Western Europe - as indicated by the first three letters of their name (in English).

In general, each country has an allocation commencing with the channel indicated and continuing to a further four channels, spaced at four channel intervals; making five channels in total. The Nordic (NOR) countries are a notable exception to this generalisation.

WARC 1977 ALLOCATIONS - PRINTED IN COUNTRY ORDER

Country	Channels Allocated	Orbit Posn. (degrees)	Boresight Lat. Long. (degrees)		Beamwidth to 3 dB points (degrees)		Centre EIRP (dBW)	Appr R.F powe (wat
Afghanistan	3,7,11,15	50	35.5	70.2	1.32	1.13	62.8	20!
Afghanistan	1,5,9,13	50	33.1	64.5	1.44	1.4	63.4	31!
Afors & Issas	21,25,29,33,37	23	11.6	42.5	.6	.6	62.5	46
Albania	22,26,30,34,38	-7	41.3	19.8	.68	.6	63.8	70
Algeria	2,6,10,14,18	-25	33.2	4.2	2.45	1.25	63.4	48!
Algeria	4,8,12,16,20	-25	25.5	1.6	3.64	2.16	62.8	10!
Andorra	4,8,12,16,20	-37	42.5	1.6	.6	.6	61.5	36
Angola	23,27,31,35,39	-13	-12	16.5	3.09	2.26	64.1	13!
Australia	3,7,11,15,19,23	98	-24.9	121.8	3.6	1.9	63	98!
Australia	1,5,9,13,17,21	98	-18.8	133.5	2.7	1.4	64.3	73!
Australia	2,6,10,14,18,22	98	-30.3	135.4	2	1.4	63.2	42!
Australia	4,8,12,16,20,24	128	-38.1	145	1.83	1.39	63.3	39!
Australia	2,6,10,14,18,22	128	-21.5	145.9	2.9	2	63.7	98!
Australia	3,7,11,15,19,23	128	-32	147.2	2.1	1.4	64.1	54!
Austria	4,8,12,16,20	-19	47.5	12.2	1.14	.63	64.1	13!
Azores	3,7,11,15,19	-31	36.1	-23.4	2.56	.7	63	25!
Bahrain	27,31,35,39	17	26.1	50.5	.6	.6	60.8	31
Bangladesh	15,18,20,22,24	74	23.6	90.3	1.46	.84	63.7	20!
Belgium	21,25,29,33,37	-19	50.6	4.6	.82	.6	64.2	93
Benin	3,7,11,15,19	-19	9.5	2.2	1.44	68	63.3	15
Botswana	2,6,10,14,18	-1	-22.2	23.3	2.13	1.5	63.7	54!
Brunei	12,14	74	4.4	114.7	.6	.6	62.5	46
Bulgaria	4,8,12,16,20	-1	43	25	1.04	.6	63.6	10!
Burma	17,19,21,23	74	19.1	97.1	3.58	1.48	63.9	94!
Burundi	22,26,30,34,38	11	-3.1	29.9	.71	.6	63.4	67
Byelorussia	21,25	23	52.6	27.8	1.08	.72	64.8	17!
Cambodia	18,20,22,24	68	12.7	105	1.01	.9	64.3	17
Cameroun	1,5,9,13,17	-13	6.2	12.7	2.54	1.68	63.4	67!
Canary Is.	23,27,31,35,39	-31	28.4	-15.7	1.54	.6	62.8	12!
Cape Verde Is.	4,8,12,16,20	-31	16	-24	.86	.7	62.2	72
Caroline Is.	1,5,9,13,17	122	8	149.5	5.36	.77	62.5	53
Central Af. Rpb	24,28,32,36,40	-13	6.3	21	2.25	1.68	64.3	73
Chad	2,6,10,14,18	-13	15.5	18.1	3.4	1.72	64	10
China	2,6,10,14	62	40.5	83.9	2.75	2.05	63.2	85!
China	1,5,9,13	62	31.5	88.3	3.38	1.45	62.9	69
China	4,8,12	62	36.3	97.8	2.56	1.58	63.5	65!
China	3,7,11	62	27.8	102.3	2.56	1.58	65.1	94
China	15,19,23	80	38	111.8	2.6	1.74	64.9	10
China	18,20,22	80	27.3	109.4	2.14	1.72	64.5	75
China	18,20,22	80	27.3	109.4	2.14	1.72	64.5	75
China	3,7,11	92	45.3	122.8	2.5	1.45	65.1	84
China	2,4,6	92	31.1	118.1	2.49	1.69	64.4	83!
China	1,5,9	92	21	115.9	2.74	2.42	63.9	11
China	1	80	39.2	116	1.2	.8	64.4	19
China	5	80	37.4	112.2	1.06	.76	64.2	15
China	9	80	41.8	111.4	1.58	1.2	63.6	31

Countries

Country	Channels Allocated	Orbit Posn. (degrees)	Boresight Lat. Long (degrees)		Beamwidth to 3 dB points (degrees)		Centre EIRP (dBW)	Approx. R.F. power (watts)
China	24	92	41.7	121.1	1.52	.78	64.5	242
China	17	92	43.7	124.3	1.98	.72	64.7	304
China	22	92	48.1	124.8	2.68	.92	65.4	619
China	16	92	36.4	118.5	1.16	.76	64.7	188
China	12	92	33	119.5	1.34	.64	64.4	171
China	10	92	32	117.2	1.2	.74	64.2	169
China	14	92	29.1	120.4	.96	.84	64.3	157
China	8	92	27.4	115.7	1.14	.94	64	194
China	15	92	25.9	118.1	1.02	.84	64.1	159
China	21	92	23.8	121.4	1.14	.82	64.3	182
China	21	80	33.9	113.7	1.2	.8	64.3	187
China	24	80	30.8	111.8	1.42	.82	64.7	248
China	12	80	27.4	111.5	1.22	.86	64.4	209
China	19	92	21.9	112.2	1.84	1.22	63.8	390
China	13	92	12.9	113.7	3.76	2.18	63.6	1360
China	14	80	23.8	108.5	1.41	1.08	64.1	283
China	17	80	35.1	108.7	1.42	.88	64.2	238
China	22	62	39	104.8	1.48	.6	63.8	154
China	20	62	37.9	101	2.?9	.82	63.7	387
China	18	62	35.4	95.7	2.1	1.14	63.4	379
China	16	62	30.2	102.5	1.91	1.23	65.5	603
China	10	80	26.7	106.6	1.14	.94	64	194
China	24	62	25.1	101.5	1.86	1.08	65	460
Comoro Is.	3,7,11,15	29	-12.1	44.1	.76	.6	63.1	67
Congo	22,26,30,34,38	-13	-.7	14.6	2.02	1.18	63.8	414
Cook Is.	2,6,10,14	158	-19.8	-161	1.02	.64	64.6	136
Cook Is. (N)	4,8,12,16	158	-11.2	-163	1.76	.72	64.3	247
Cyprus	21,25,29,33,37	5	35.1	33.1	.6	.6	63.6	59
Czechoslavakia	3,7,11,15,19	-1	49.3	17.3	1.47	.6	63.8	153
Denmark	12,16,20	5	57.1	12.3	1.2	.6	64.3	140
Denmark	24,36	5	61.5	17	2	1	67.5	814
Denmark	27,35	5	61	-19.5	2.2	.8	66.2	531
East Germany	21,25,29,33,37	-1	52.1	12.6	.83	.63	64.2	99
Egypt	4,8,12,16,20	-7	26.8	29.7	2.33	1.72	63.1	592
Eq. Guinea	23,27,31,35,39	-19	10.3	10.3	.68	.6	63.8	70
Ethiopia	22,26,30,34,38	23	9.1	39.7	3.5	2.4	63.4	1331
Fiji	1,5,9	152	-17.9	179.4	1.04	.98	63.7	173
Finland	2,6,10	5	64.5	22.5	1.38	.76	67.7	447
Finland	22,26	5	61.5	17	2	1	67.7	853
France	1,5,9,13,17	-19	45.9	2.6	2.5	.98	63.8	425
Gabon	3,7,11,15,19	-13	-.6	11.8	1.43	1.12	63.3	248
Gambia	3,7,11,15,19	-37	13.4	-15.1	.79	.6	63.3	73
Ghana	23,27,31,35,39	-25	7.9	-1.2	1.48	1.06	63.6	260
Greece	3,7,11,15,19	5	38.2	24.7	1.78	.98	63.3	270
Guam	2,6,10,14,18	122	13.1	144.5	.6	.6	63.3	55
Guinea	1,5,9,13,17	-37	10.2	-11	1.58	1.04	63.4	260

Countries

Country	Channels Allocated	Orbit Posn. (degrees)	Boresight Lat. Long (degrees)	Beamwidth to 3 dB points (degrees)		Centre EIRP (dBW)	Approx R.F. power (watts)
Guinea-Bissaw	2,6,10,14,18	-31	12	-15	.9	.6	63.1	79
Hungary	22,26,30,34,38	-1	47.2	19.5	.92	.6	64	100
Iceland	21,25,29,33,37	-31	64.9	-19	1	.6	65.8	165
Iceland	23,31,39	5	61	-19.5	2.2	.8	66.3	543
India	2,6,10,14	68	25.5	93	1.46	1.13	63.9	293
India	17,19,21,23	56	33.4	75.9	1.52	1.08	64.3	320
India	1,5,9,13	56	11.2	72.7	1.26	.6	63.1	111
India	4,8,12,16	56	25	73	1.82	1.48	63.6	447
India	18,20,22,24	56	16	78.4	2.08	1.38	63.8	498
India	18,20,22,24	68	27.7	79.3	2.14	1.16	63.8	431
India	3,7,11,15	56	11.1	77.8	1.36	1.28	63.3	269
India	1,5,9,13	68	22.3	79.5	2.19	1.42	63.3	481
India	2,6,10,14	56	19.5	76.2	1.58	1.58	63.5	404
India	17,19,21,23	68	20.5	84.7	1.6	.86	63.6	228
India	3,7,11,15	68	11.1	93.3	1.92	.6	63.4	182
India	4,8,12,16	68	25	86.2	1.56	.9	63.7	238
Indonesia	2,4,6,8	80	0	101.5	3	1.2	63.3	557
Indonesia	18,20,22,24	80	-8.1	112.3	3.14	1.46	64.2	873
Indonesia	17,19,21,23	80	-.3	112.3	2.66	2.32	64	1122
Indonesia	1,5,9,13	104	-3.2	124.3	3.34	1.94	63.2	980
Indonesia	3,7,11,15,19	104	-3.8	135.2	2.46	2	63.8	854
Iran	3,7,11,15,19	34	32.4	54.2	3.82	1.82	62.8	959
Iraq	24,28,32,36,40	11	32.8	43.6	1.88	.96	63.3	279
Ireland	2,6,10,14,18	-31	53.2	-8.2	.84	.6	64.2	96
Israel	25,29,33,37	-13	31.4	34.9	.94	.6	63.8	98
Italy	24,28,32,36,40	-19	41.3	12.3	2.38	.98	64.1	434
Ivory Coast	22,26,30,34,38	-31	7.5	-5.6	1.6	1.22	63.7	331
Japan	1,3,5,7,9,11,13,15	110	31.5	134.5	3.52	3.3	63.2	1758
Jordan	23,27,31,35,39	11	31.4	35.8	.84	.78	63.1	96
Kenya	21,25,29,33,37	11	1.1	37.9	2.29	1.56	63.7	606
Kuwait	22,26,30,34,38	17	29.2	47.6	.68	.6	63.1	60
Laos	2,4,6,8,10	74	18.1	103.7	2.16	.78	63.8	292
Lebanon	3,7,11,15,19	11	33.9	35.8	.6	.6	61.6	37
Lesotho	24,28,32,36,40	5	-29.8	27.8	.66	.6	64.2	75
Liberia	3,7,11,15	-31	6.6	-9.3	1.22	.7	63.2	129
Libya	1,5,9,13,17	-25	26	21.4	2.5	1.04	63.5	421
Libya	3,7,11,15,19	-25	27.2	13.1	2.36	1.12	63	382
Liechtenstein	3,7,11,15,19	-37	47.1	9.5	.6	.6	62.4	45
Luxembourg	3,7,11,15,19	-19	49.8	6	.6	.6	62.9	50
Malagasy	1,5,9,13,17	29	-18.8	46.6	2.72	1.14	63.3	480
Malawi	24,28,32,36,40	-1	-13	34.1	1.54	.6	64.2	176
Malaysia	16,18,20,22,24	86	4.1	102.1	1.62	.82	63.2	201
Malaysia	2,4,6,8	86	3.9	114.1	2.34	1.12	63.6	434
Maldives	12,16	44	6	73.1	.96	.6	63.7	97
Mali	2,6,10,14,18	-37	19	-2	2.66	1.26	63.2	507
Mali	4,8,12,16,20	-37	13.2	-7.6	1.74	1.24	63.7	366

Countries

Country	Channels Allocated	Orbit Posn. (degrees)	Boresight Lat. (degrees)	Long (degrees)	Beamwidth to 3 dB points (degrees)		Centre EIRP (dBW)	Approx. R.F. power (watts)
Malta	4,8,12,16	-13	35.9	14.3	.6	.6	61	32
Marianas Is.	3,7,11,15,19	122	16.9	145.9	1.2	.6	63.5	116
Marshall Is.	2,6,10,14,18	146	7.9	166.7	1.5	1.5	63.3	348
Mauritania	22,26,30,34,38	-37	18.5	-12.2	2.62	1.87	62.8	676
Mauritania	24,28,32,36,40	-37	23.4	-7.8	1.63	1.1	63	259
Mauritius	2,6,10,14,18	29	-18.9	59.8	1.62	1.24	64	365
Mauritius	4,8,12,16	29	-13.9	56.8	1.56	1.38	63.7	365
Mayotte	24,28,32,36,40	29	-12.8	45.1	.6	.6	63.4	57
Monaco	21,25,29,33,37	-37	43.7	7.4	.6	.6	62.4	45
Mongolia	25,29,33,37,39	74	46.6	102.2	3.6	1.13	64.1	757
Morocco	21,25,29,33,37	-25	29.2	-9	2.72	1.47	63.3	619
Mozambique	4,8,12,16,20	-1	-18	34	3.57	1.38	64.2	938
Namibia	25,29,33,37	-19	-21.6	17.5	2.66	1.9	64.7	1080
Nauru	3,7,11,15	134	-.5	167	.6	.6	62.5	46
Nepal	17,19,21	50	28.3	83.7	1.72	.6	64.6	215
Netherlands	23,27,31,35,39	-19	52	5.4	.76	.6	64.4	90
New Caledonia	2,6,10,14	140	-21	166	1.14	.72	63.7	139
New Hebrides	3,7,11,15	140	-16.4	168	1.52	.68	62.8	142
New Zealand	1,5,9,13	158	-39.7	172.3	2.88	1.56	63.3	695
New Zealand	13,17,21	128	-41	173	3.3	1.28	64.8	924
Niger	24,28,32,36,40	-25	16.8	8.3	2.54	2.08	64.5	1078
Nigeria	22,26,30,34,38	-19	9.4	7.8	2.16	2.02	63.9	775
Niue	19,23	158	-19	-169.8	.6	.6	64.1	67
North Korea	14,16,18,20,22	110	39.1	127	1.3	1.1	64	260
Norway	14,18,38	5	64.1	13.1	1.84	.88	65	370
Norway	28,32	5	61.5	17	2	1	66.8	693
Oceania	4,8,12,16	-160	-16.3	-145	4.34	3.54	63.5	2491
Oman	24,28,32,36,40	17	21	55.6	1.88	1.02	63.3	297
Pakistan	2,6,10	38	29.5	69.6	2.3	2.16	63.9	883
Pakistan	12,14	38	30.8	72.1	1.16	.72	63.5	135
Pakistan	18,22	38	27.9	65.2	1.52	1.42	63	311
Pakistan	20,24	38	25.8	68.5	1.32	.62	63.3	126
Pakistan	4,8	38	33.9	74.7	1.34	1.13	64.3	295
Palmyra	1,5,9,13,17	170	7	-161.4	.6	.6	62.4	45
Papua N. Guinea	2,6,10,14	110	-6.3	147.7	2.5	2.18	64.4	1087
Papua N. Guinea	4,8,12	128	-6.7	148	2.8	2.05	64.3	909
Philippines	16,18,20,22,24	98	11.1	121.3	3.46	1.76	63.7	1034
Poland	1,5,9,13,17	-1	51.8	19.3	1.46	.64	64.1	173
Portugal	3,7,11,15,19	-31	39.6	-8	.92	.6	63.4	87
Qatar	1,5,9,13,17	17	25.3	51.1	.6	.6	61.8	39
Reunion	22,26,30,34,38	29	-19.2	55.6	1.56	.78	63.9	216
Romania	2,6,10,14,18	-1	45.7	25	1.38	.66	63.8	158
Rwanda	4,8,12,16,20	11	-2.1	30	.66	.6	64.8	86
Samoa	1,5,9,13,17	170	-14.2	-170.1	.6	.6	61.1	33
San Marino	1,5,9,13,17	-37	43.7	12.6	.6	.6	62.4	45
Sao Tome	4,8,12,16,20	-13	.8	7	.6	.6	61.4	35

Countries

Country	Channels Allocated	Orbit Posn. (degrees)	Boresight Lat. Long. (degrees)		Beamwidth to 3 dB points (degrees)		Centre EIRP (dBW)	App R. pow (wa
Saudi Arabia	4,8,12,16,20	17	23.8	41.1	3.52	1.68	62.7	79
Saudi Arabia	2,6,10,14,18	17	24.6	48.3	3.84	1.2	62.7	62
Senegal	21,25,29,33,37	-37	13.8	-14.4	1.46	1.04	63.6	25
Sierra Leone	23,27,31,35,39	-31	8.6	-11.8	.78	.68	63.4	84
Singapore	3,7,11,15	74	1.3	103.8	.6	.6	63.5	58
Somalia	3,7,11,15,19	23	6.4	45	3.26	1.54	62.3	61
South Africa	21,25,29,33,37	5	-28	24.5	3.13	1.68	64.1	97
South Korea	2,4,6,8,10,12	110	36	127.5	1.24	1.02	63.6	20
South Yemen	1,5,9,13,17	11	15.2	48.8	1.76	1.54	62.8	37
Spain	23,27,31,35,39	-31	39.9	-3.1	2.1	1.14	63.9	42
Sri Lanka	2,6,10,14	50	7.7	80.6	1.18	.6	63.6	11
Sudan	23,27,31,35,39	-7	7.5	29.2	2.34	1.12	64.4	52
Sudan	22,26,30,34,38	-7	12.7	28.9	2.26	1.96	63.5	71
Sudan	24,28,32,36,40	-7	19	30.4	2.44	1.52	63.3	57
Swaziland	1,5,9,13,17	-1	-26.5	31.5	.62	.6	62.8	51
Sweden	4,8,34	5	61	16.2	1.04	.98	67.1	37
Sweden	30,40	5	61.5	17	2	1	67.1	74
Switzerland	22,26,30,34,38	-19	46.6	J.2	.98	.7	64.1	12
Syria	22,26,30,34	11	34.9	38.3	1.04	.9	63.2	14
Syria	38	11	34.2	37.6	1.32	.88	63.4	18
Tanzania	23,27,31,35,39	11	-6.2	34.6	2.41	1.72	63.7	70
Thailand	1,5,9,13	74	13.2	100.7	2.82	1.54	63.6	72
Togo	2,6,10,14,18	-25	8.6	.8	1.52	.6	63.4	14
Tokelau Is.	20,24	158	-8.9	-171.8	.7	.6	63.8	72
Tonga	4,8,12,16	170	-18	-174.7	1.41	.68	63.3	14
Tunisia	22,26,30,34	-25	33.5	9.5	1.88	.72	63.8	23
Tunisia	38	-25	32	2.5	3.59	1.75	61.9	70
Turkey	1,5,9,13,17	5	38.9	34.4	2.68	1.04	63.7	47
Uganda	3,7,11,15,19	11	1.2	32.3	1.46	1.12	63.2	24
Ukraine	29,33,37	23	48.4	31.2	2.32	.96	64.6	46
United Kingdom	4,8,12,16,20	-31	53.8	-3.5	1.84	.72	65	30
Upper Volta	21,25,29,33,37	-31	12.2	-1.5	1.45	1.14	64	30
U.S.S.R.	27,31,35,39	23	47	36	3.7	1.43	65.2	12
U.S.S.R.	4,8,12,16	23	57.4	41.5	3.08	1.56	66.7	16
U.S.S.R.	3,7,11,15,19,23	23	56.6	24.7	.88	.64	65	12
U.S.S.R.	1,5,9,13,17,23	23	40.8	45.6	2.16	.6	63.9	23
U.S.S.R.	20	23	63.1	32.4	1.18	.6	66.6	23
U.S.S.R.	20,24,28,32,36,40	44	44.6	64.3	4.56	2.48	65.4	28
U.S.S.R.	1,5,9,13	44	58.5	62.4	3.2	1.52	66.3	1
U.S.S.R.	26,30	44	38.8	59	2.24	1	64	40
U.S.S.R.	12,16	44	38.5	70.8	1.36	.74	64.1	18
U.S.S.R.	18,22	44	41	73.9	1.34	.84	64.5	22
U.S.S.R.	34,38	44	63.3	42	2.64	.84	64.4	44
U.S.S.R.	7	44	61.5	70.1	2.38	.66	67.1	58
U.S.S.R.	3	44	63.5	54.3	1.58	.66	66.9	36
U.S.S.R.	26,30,34,38	74	57.6	88.8	3.08	1.68	67.9	23

Countries -

Country	Channels Allocated	Orbit Posn. (degrees)	Boresight Lat. Long. (degrees)		Beamwidth to 3 dB points (degrees)		Centre EIRP (dBW)	Approx. R.F. power (watts)
U.S.S.R.	32	74	51.7	94	1.52	.6	65.1	213
U.S.S.R.	28	74	63.2	98	1.84	.69	68.1	593
U.S.S.R.	19,23,27,31,35,39	110	57.3	112.7	2.67	1.75	64.1	870
U.S.S.R.	25	110	53.4	108.2	2.16	.78	65	385
U.S.S.R.	20,24,28,32,36,40	140	53.6	138	3.16	2.12	67.7	2857
U.S.S.R.	26,30,34,38	140	55.4	155.3	2.9	2.36	67.9	3057
U.S.S.R.	22	140	65.5	168.5	1.96	.6	68.1	550
U. A. Emirates	21,25,29,33,37	17	24.2	53.6	.98	.8	63.2	118
Vatican	27,31,35,39	-37	41.8	12.4	.6	.6	65.2	86
Vatican	23	-37	41.5	10.8	2	.6	63.6	199
Vietnam	3,7,11,15	86	16.1	105.3	3.03	1.4	63.4	672
Wake Is.	1,5,9,13,17	140	19.2	166.5	.6	.6	63.6	59
Wallis Is.	2,6,10,14	140	-14	-176.8	.74	.6	64.4	88
Western Samoa	3,7,11,15	158	-13.7	-172.3	.6	.6	63.6	59
West Germany	2,6,10,14,18	-19	49.9	9.6	1.62	.72	65.5	299
Yemen	2,6,10,14,18	11	15.1	44.3	1.14	.7	62.6	105
Yugoslavia	21,25,29,33,37	-7	43.7	18.4	1.68	.66	65.2	265
Yugoslavia	23,27,31,35,39	-7	43.7	18.4	1.68	.66	65.2	265
Zaire	4,8,12,16,20	-19	0	22.4	2.16	1.88	64.7	868
Zaire	2,6,10,14,18	-19	-6.8	21.3	2.8	1.52	64.6	889
Zambia	3,7,11,15,19	-1	-13.1	27.5	2.38	1.48	63.7	598
Zimbabwe	22,26,30,34,38	-1	-18.8	29.6	1.46	1.36	64.2	378

Notes on the WARC 1977 tables

1. The information contained in these tables is derived from the planning document generated by the World Administrative Radio Conference in 1977

2. The country name has been derived from the IFRB number given in the documents interpreted to be the commonly used name at the time of writing

3. Channel numbers, orbital positions, boresight, beamwidth and centre EIRP powers are as in the WARC documents

4. Throughout the tables negative quantities are used for degrees latitude West or longitude South

5. The approximate radio frequency power required from the transmitting device in the satellite is calculated from the centre EIRP and the beamwidths using the formula:-

    $$\text{Power (dBW)} = \text{EIRP} + 3 - (44.4 - \log(a) - \log(b))$$

    and:-

    $$\text{Power (watts)} = 10^{(\text{Power (dbW}/10))}$$

    Where the half-power beamwidths of the beam are a and b

    The expression '44.4-log(a)-log(b)' is that recommended by the CCIR as a way of calculating the centre gain (in dBi) of an antenna. The factor of 3 dB is inserted to allow for losses between transmitting device and antenna in the satellite and antenna losses.

6. The approximate power required is only appropriate if the satellite is designed to cover the service area with the maximum flux-density permitted in the WARC '77 plan. If a lower EIRP is used or if a shaped beam antenna is employed to direct the radio energy in a selective manner to the service area then smaller transmitter powers can be used.

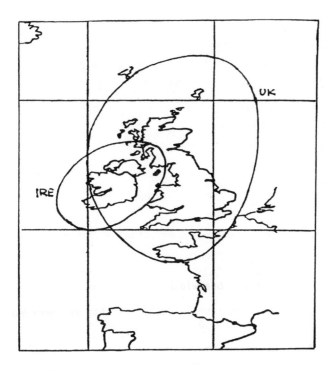

Fig.4.5 Footprint based on WARC Plan

4.5 COVERAGE MAPS

4.5.1. The Footprints

The coverage area for any country, as provided in the plan, has a near elliptical shape. This is obtained using a relatively simple satellite transmitting antenna. Although more beam shaping is technically possible this was largely ignored in the plan.

The antenna performance outside the coverage area depends on its design details. In that the performance should be better than the prescribed template the diagrams Fig 4.7 and 4.8 will be optimistic.

A practical minimum size of an axis of the coverage

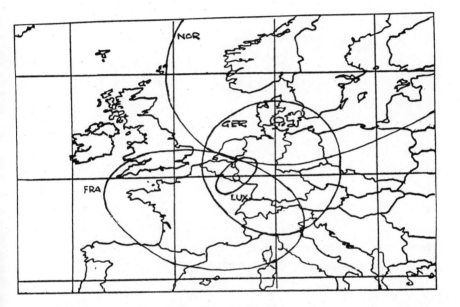

Fig. 4.6 Footprint based on WARC Plan

ellipse was considered to be 0.6 degrees. Thus small
countries inevitably received relatively large coverage
assignments.

The coverage area is often described as the "footprint".
Examples of the footprint for several major countries in
Europe are as detailed in Fig.4.5 and Fig.4.6.

4.5.2. Overspill

Although the footprints are shown as areas defined by
firm lines they are in fact imprecise. The radiated
power does not cease abruptly but tapers off as one
moves away from the intended target area. However if
a somewhat larger receive antenna is used then a good
signal can still be received.

Thus the signal spills over the neighbouring countries
and this "overspill" is important. It could be a
significant source of additional programmes for domestic
or cable use. However it brings with it many problems.
There are legal problems, political problems and problems
of programme rights and the associated finance.

The spillover problem has been made much greater by the
advance of technology. Receivers can now be made much
better than that anticipated by the WARC and, in
situations where the quality of picture is limited only
by noise, it will be possible to receive good pictures
over much wider areas. This change is particularly
important because it would appear to be many years before
all the assignments are taken up and Europe is interference limited.

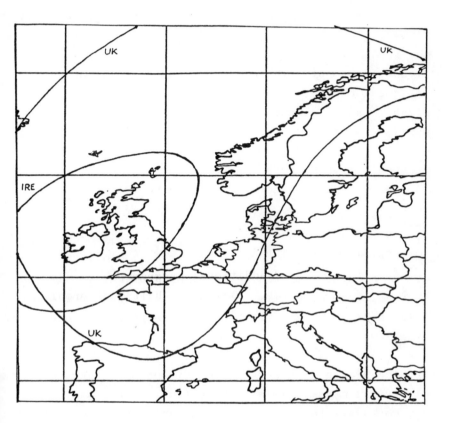

Fig. 4.7 Theoretical footprint based on the use of
90 cm dish antenna and a receiver employing
1983 technology (signal not interference
limited)

The 1977 receiver with about 8 dB Noise Figure had now been replaced by one of about 2.5 dB and, in an interference free situation, this is equivalent to an increase of transmitter power by about 400% above that intended by the WARC.

The resulting footprints are illustrated in Fig.4.7 and Fig.4.8.

It should be noted the United Nations Resolution (see the legal chapter) prohibits the broadcasting of a signal overtly intended for another country without the permission of that country.

It should also be noted that in order to receive an "overspill" signal the viewer may have to:

    a) retune his receiver

    b) adjust his aerial (slew)

    c) alter the signal polarisation

At the present time the receiver will probably only tune over half the band ,i.e. it operates in the upper or lower half. (This limitation may disappear as technology progresses).

Antenna slewing may be achieved by mechanical or electrical means. It will be noted that Britain, Spain and Portugal are at position -31° and France, Germany, Switzerland, Belgium and Netherlands are on position -19° and so slewing will be a requirement for many people.

A polarisation switch will add to the cost of the receive aerial array.

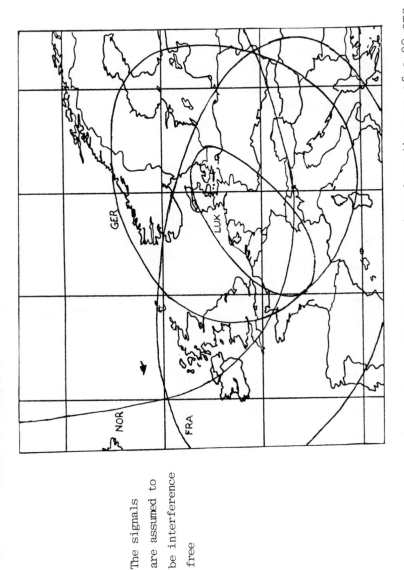

The signals
are assumed to
be interference
free

Fig 4.8 Theoretical footprint based on the use of a 90 cms
dish and a receiver using 1983 technology

4.6 THE DIFFERENCES BETWEEN THE WARC AND THE REGIONAL ADMINISTRATIVE RADIO CONFERENCE (RARC REGION 2)

4.6.1. Introduction

In the five years between these two conferences there had been many changes.

There have been significant advances in receiver design and a low noise front end for the receiver is now a commercial proposition. As a result the modern DBS receiver can operate with a lower signal input.

Secondly, satellite broadcasting was now seen to be a real opportunity and there were strong pressures laying claim to the limited frequency spectrum available for broadcasting. Experiment and study showed how to make better use of the spectrum and it was possible to provide more services.

Last, but not least, there were growing ambitions to introduce a new service of high definition television. The allocation plan chosen was such as to make possible the grouping of adjacent channels so as to provide a sufficient bandwidth to support HDTV.

The RARC in 1983 planned the overall system of satellite broadcasting whereas the WARC in 1977 had concentrated its attention principally on the downlink from satellite to viewer. Both television feeding uplinks and broadcasting downlinks are included in the Region 2 plan and this attention to a wider range of problems led to several differences between the two plans.

4.6.2. Orbital positions

The RARC decided to use orbital
positions 9° apart and to locate at each position two
clusters of satellites 0.4 degrees apart. This
separation of 0.4 degrees has no effect on the downlink
operation because the beamwidth of domestic receivers
is sufficient to accept signals from satellites in
either of two groups. However, for the uplink case
where larger earthstation antennae are used, the
separation provides a useful additional interference
margin between adjacent frequency channels. As any one
broadcaster is likely to be allocated channels on only
one polarisation the 0.4 degree orbit separation is
unlikely to lead to provision of more uplink earthstations.

As a result of this innovation more of the available
frequencies can be used and the uplink congestion has
been eased.

There are, in general, two orbital positions for each
time zone (one overhead and one 50° West).

On each position 32 frequency allocations are used
(16 with one polarisation, 16 with the alternative
polarisation).

NORTH AMERICAN ORBITAL POSITIONS

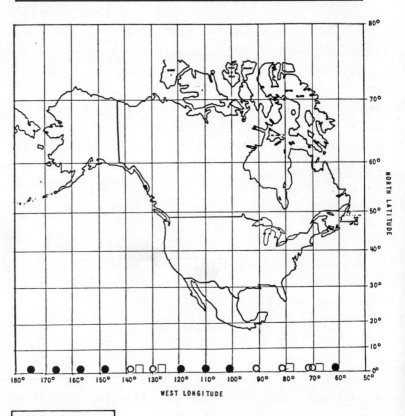

Fig. 4.9

Shuttle (STS) on launch pad 39A, Kennedy Space Centre (Courtesy of European Space Agency)

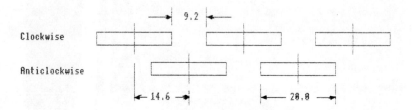

Fig. 4.10   Channel Separation - MHz

RARC Plan

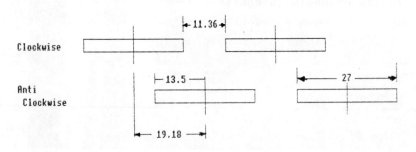

Fig. 4.11   Channel Separation - MHz

WARC Plan

4.6.3. The Frequency Plan

Although the two Regions had different approaches the results were similar and the plan adopted by these two conferences is illustrated in Fig.4.10 and Fig.4.11. The bandwidth of each channel may not be abruptly limited as shown but it extends beyond these boundaries in a manner dependant on the modulation index.

In discussing the channel packing techniques used in drawing up the RARC '83 Plan it should be noted that only 500 MHz of spectrum was made available in comparison with the 800 MHz in Region 1.

The two conferences also decided to allocate the channels in a different manner. In general the WARC chose to allocate channels to each country spaced at four channel intervals. The RARC however decided to allocate the channels in contiguous blocks, believing that this would open the door to further developments and east adjacent channel interference in the international scene.

There has been a marginal increase in the modulation index and the interference specification has been slightly relaxed. However, many would say that the geographical constraints in the three regions are different and interference will be less of a problem in Region 2.

Region 1 was predominantly concerned with the 625 line
television systems (PAL/SECAM) and their derivatives.
These had a maximum bandwidth of 6.0 MHz.

Region 2, on the other hand, had to provide for the
NTSC television system with a bandwidth of 4.2 MHz.
Naturally the total requirement had to include provision
for the sound system and the plans were based on 6.5 MHz
baseband for Region 1 (PAL/SECAM) and 4.5 MHz baseband
for Region 2 (NTSC). The channel size and separation
in the two regions are approximately in this ratio.
(This is not to say that wider bandwidths will not be
used on a non interference basis).

4.6.4. Power Flux Density

The other major difference was the limiting power flux
density at the edge of the band. In Region 2 a power
flux density of -107 dB W/M^2 was proposed (Receiver G/T
10dB/K). The USA reserved the right to use -105dHW/M^2
if this became necessary. By comparison it will be
recalled that Region 1 adopted a figure of -103 dbW at
the edge of the footprint.

The difference in power and channel spacing causes
problems in the overlap areas between Region 1 and
Region 2 which required special legislation.

4.6.5. Other Downlink Features

There are downlink differences between the two plans
in addition to the different channel bandwidths and
EIRPs discussed above.

In particular the downlink radiation patterns for co-
and cross-polarisation are both significantly different.
It is said that they were drawn up taking note of the
practical difficulties of achieving the Region 1
patterns and to allow for the use of shaped beam down-
link patterns - which in general generate a higher
sidelobe level than simple elliptical pattern antennas.
Shaped beam antennas using oversize reflectors and
multiple feeds can achieve WARC '77 sidelobe levels
(TDF-1 uses such an antenna) but require the use of
sidelobe-suppressing feeds which add to the mass and
complexity of the antenna. The RARC '83 patterns do
not appear to require special sidelobe suppression
techniques. The USA satellite broadcasters are planning
to use shaped beam antenna coverage patterns tailored to
direct energy just at the coverage area required.

DBS plans for Regions 1 and 3 compared with those for Region 2

	Regions 1 & 3	Region 2
1. Orbit spacing (note 1)	6 degrees	9 degrees
2. Channel spacing	19.18 MHz	14.58 MHz
3. Channel bandwidth	27 MHz	20 MHz
4. Number of channels (note 2)	40	32
5. Energy dispersal	600 KHz	none
6. Power flux density (note 3) (W/m^2)	-103 dB	-107 dB
7. Receiver G/T (dB/K)	6	10
8. Antenna size	0.9 metre	1.0 metre
9. Protection ratios		
- co-channel	31 dB	29 dB
- adjacent channel	15 dB	13.6 dB

Notes

1. Region 2 allocation in pairs at each orbital position

2. Signals on each frequency may have one of two alternative polarisations

3. For 99% of the worst month

Region 2

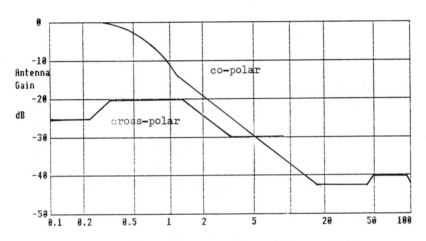

Fig. 4.13   Reference patterns for receiving
earth station antenna

Relative $\left(\dfrac{\varnothing}{\varnothing_o} \right)$
angle

Region 2

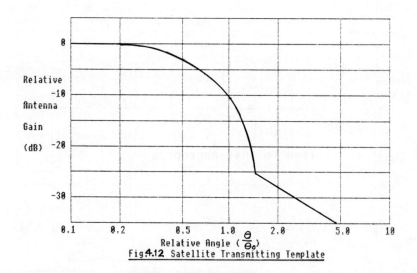

Fig 4.12 Satellite Transmitting Template

4.7 OPPORTUNITIES AT OTHER FREQUENCIES

The 1971 conference recognised the possibility of
broadcasting at other frequencies however they all have
disadvantages of one sort or another and they have only
been considered by a few countries.

4.7.1. Sound broadcasting in the band 0.5 to 2.0 GHz

Studies are recommended by all interested Administrations
on the possibilities of broadcasting in the band 1.429 to
1.525 GHz. Consideration will be given to this possibility at the next frequency conference.

4.7.2. Satellite broadcasting in the band 2.500 to 2.690

All three Regions have provisions for use of this band
on a shared basis. Each region has a number of different
reservations but all agree that, in this band, the power
should be limited to that appropriate for community
reception.

4.7.3. Satellite broadcasting in the 11 - 12 GHz band

Region 1 ..... as detailed previously 11.7 to 12.5 GHz
Region 2 ..... 12.1 to 12.7 GHz (subject to confirmation)
Region 3 ..... 11.7 - 12.75 GHz

In Region 2 the frequency band adjacent to the satellite
broadcasting band, 11.7 - 12.2 GHz, is available for
moderate power satellites of up to 53 dBW/tv channel EIRP
- some 10 dBs below full direct broadcasting power. In
addition the DBS band itself, 12.2 - 12.7 GHz is
available also for telecommunication purposes. The
planners inserted these provisions to allow for the
flexible use of frequencies in Region 2 and encourage the

emergence of new types of satellite services. Such flexibility also prevents the sterilisation of bands of frequencies in the event that the services planned for them do not materialise.

4.7.4. Satellite broadcasting at 22 GHz

There is an allocation in Regions 2 and 3 between 22.5 and 23.00 GHz on a shared basis. (This band is allocated on a primary basis in Japan).

4.7.5. Satellite broadcasting in the band 40.5 to 42.5GHz

There is a shared allocation for all Regions in this band. However these frequencies suffer from severe precipitation attenuation (see chapter 3).

4.7.6. Satellite broadcasting in the band 84 - 86 GHz

There is a shared allocation for all Regions in this band. As above these frequencies are also subject to severe attenuation.

NOTE: The priorities for the allocation of frequencies to competing services is different in each Region. All these broadcasting bands are affected by a number of minor reservations and reference should be made to the Final Acts of the World Administrative Conference in 1977 if further details are required.

It should be noted that an Administration does not have to make allocations in these bands.

Furthermore although these bands permit shared services an Administration may decide to allocate it solely for use by one of the services.

4.8.BASEBAND STANDARDS FOR DBS

The design of a DBS system and result obtained is very
dependent on the technical standards employed. Many
factors influence the choice of standards. Technical
excellence is, of course, important but so is receiver
cost, the development time and the effect of the new
system on the receiver industry in each country. Allowance must also be made for the effect on competitive
services such as cable and the compatibility with
existing receivers.

The choice could be to use terrestial standards or to
devise new and more suitable standards and there are many
alternatives.

Those in the debate divide naturally into three groups.
First there are those who regret the existence of many
different television standards, particularly in Europe,
and seek to establish one common standard. The standard
most likely to bring about unity was thought to be a new
standard which was flexible enough to allow development
in the future. As an example the C-MAC standard is
flexible and does offer improved picture quality. There
has been widespread acceptance of this system by engineers in Europe. Time will tell whether the system is
successful in winning over the governments of Europe.

Secondly there is a group who see the growth of a new
service to be limited by financial constraints. They
seek the minimum cost receiver and consider the launch
of the new service must be associated with the use of
converters and existing receivers and VCR's. It is
likely that terrestial television will develop a digital

stereo sound system. In these circumstances this group considers the best solution to be the compatible use of these improved terrestrial television standards on the new DBS service.

Some see this development as a temporary solution with the system developing to better standards as the years go by; others see it as a more permanent solution.

Thirdly there is a group that seeks to introduce high definition television. They want a standard which is a significant step forward to a wide aspect ratio picture with higher horizontal and vertical resolution than is offered by any system in use today.

The characteristics of these three groups will now be considered in terms of the WARC Plan and from the point of view of those intending to launch a new business. Thus the descriptions of the various proposals are intended to be sufficient to enable estimates to be made of the power or signal to noise ratio requirements and the resulting picture quality. They also outline any new facilities these standards have to offer because the market may find these new facilities attractive and, accordingly, it will improve the viability of a DBS service.

Reference should be made to the current literature if further details are required. Unfortunately it is not possible to do full justice to these systems within these few pages.

4.8.1. First, 625 Line Colour Component System

In Europe many felt that there was an opportunity to introduce an improved standard that offered the prospect

of better picture and sound services. The concept was
to create a standard which took advantage of the new
integrated circuit technology. Hopefully, using this
technology, technical complication in the receiver would
not cause an escalation in its cost. Under the auspices
of the EBU (European Broadcasting Union) a new system
was developed using the colour components (the colour
difference signals R-Y (Red - Luminance) and B-Y (Blue-
Luminance)). The system employed allowed a near perfect
decoding at the receiver. It offered improved resolution
and a picture with less artifacts. It also gave better
stereo sound quality and it offered the possibility of
broadcasting a large number of independent sound and
data services. It combined these relatively new ideas
on vision signal coding with the new concept of digital
sound transmitted in "packets".

The EBU "MAC" system (Multiplexed Analogue Components)
sends the two components, one after the other, that is
in time division multiplex. The sound is also sent
in time division multiplex known as a packet "C"
system. Thus the complete package is known to many as
the "C-MAC" system (see Appendix).

The performance details given in Fig.4.14 are the
results of an experiment which sought to illustrate
the degradation in picture quality resulting from all
forms of impairment. Although the results are very
similar to those quoted in Chapter 5 the information
given in that chapter refers only to noise impairment
and is better substantiated. When making calculations
about the signal-to-noise ratio performance and the
system power levels reference should be made to that
chapter.

The authors wish to take this opportunity to acknowledge
the pioneering work done by the Independent Broadcasting
Authority on the MAC television system.

4.8.2. Second, 625/525 Line Composite Systems

All picture coding systems have to employ a multiplex
system in order to transmit both the luminance and
chrominance signals in the same channel. NTSC, PAL and
SECAM all use frequency division multiplexing, where
the colouring signal is modulated on a high frequency
carrier. NTSC and PAL use amplitude modulation and the
SECAM system uses frequency modulation of this carrier.
These are known as composite systems because the
luminance and chrominance signals are combined and sent
simultaneously as one signal.

The principal characteristics of these standards are
listed in Fig.4.15.

The properties of these systems are well known and they
set a standard by which all other systems are judged.

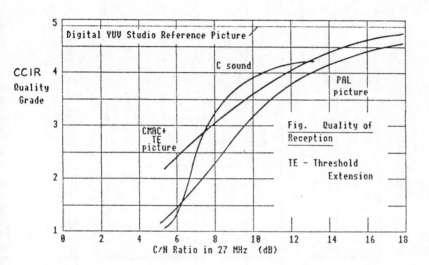

Fig. 4.14   System Performance

These results are based on the work of members
of the European Broadcasting Union.

(see Appendix 10)

Note, the sound system may be improved by adding
further error correction

Origin /Name 1.	Lines/ Picture 2.	Picture Rate 3. /S	4.Bandwidth (MHz) Luminance	4.Bandwidth (MHz) Total System	Aspect Ratio 5.	Column Ref.	Notes
NTSC	525	30	4.2	4.5	4:3	4	Colour components sent as Frequency Division Multiplex (FDM) with sub-carrier at 3.53 MHz.
PAL	625	25	5.5	6.0	4:3	4 1	As above but with sub-carrier at 4.33 MHz. System I
SECAM	625	25	6.0	6.5	4:3	4	Colour components Frequency modulated on 4.3 MHz sub-carrier with amplitude modulated pre-emphasis.
C-MAC	625	25	5.5	8.25	4:3	4	Time Division Multiplex of Luminance, colour difference and sound signals. Vision signals time compressed for transmission.

Fig 4.15   Existing Terrestial Standards compared with C-MAC
(There are many variants of these three standards)

Origin /Name 1.	Lines/ Picture 2.	Picture Rate 3.	4.Bandwidth (MHz) Luminance	Total System	Aspect Ratio 5.	Column Ref.	Notes
MUSE (NHK)	1125	30	20+	8	5:3	1 3 4 2	A Transmission standard On Moving Pictures Resolution on stationary pictures. Interlaced picture
C-MAC (Ext)	625	25	8	12	4.7:3	2/3 4	Anticipates receiver synthesis of 625 line sequential pictures at 100Hz repetition rate. Time Division Multiplex including sound and vision components. A compatible extension of the C-MAC system

Fig.4.16 Some proposed High Definition Television Systems.

4.8.3. Thirdly, High Definition Television Standards

The size and quality of the television display has
always had a major influence on the television broadcast
standards. Early monochrome standards were satisfactory
for many purposes and colour television standards were
only introduced when a practical colour television tube
was available. In the more recent years we have seen
the size of these television displays growing larger
and larger. It is now clear that it will not be long
before the limitations of the existing television
standards will be clear for all to see. However true
large screen HDTV will not be possible before a new
display has been devised. At this date it would be rash
to guess whether this will be a projection receiver or
a new form of cathode ray tube with a flat screen.
Perhaps it will be an entirely different technology?

There seems to be little doubt that it will employ a
different aspect ratio display and offer wide screen
pictures. If existing line standards are employed we
will find the line structure obtrusive. In addition
large bright pictures will accentuate existing flicker
problems. Thus the new standard will have to have to
significantly increase line number and an increase in
picture/field rate. There are those who suggest that
the new standard must offer something dramatically
better if the public are to be enticed to buy the new
equipment. Perhaps this will be the opportunity to
introduce a worldwide standard! Clearly it will be
some time before these ambitions are realised but
opinions differ as to the precise date in the future
when this will be possible.

The bandwidth restriction forces the HDTV system development to divide into three segments. The studio must originate programmes on a standard whose technical quality enables it to penetrate any market. The transmission system must modify or adapt this signal to limit the bandwidth requirement and the receiver must then process this signal so as to reconstruct and display the HDTV picture. Figures 4.16 and 4.17 give some examples of proposed high definition standards.

Clearly the search to find a new HDTV transmission standard suitable for broadcasting in the 12 GHz band will continue. If this signal can be downwards compatible with existing receivers then launch of the new service will be that much easier (see C-MAC extended, Fig.4.16).

Region 2 also has the opportunity to introduce HDTV in the 22 GHz band but no such provision has been made in Region 1.

Features influencing the design of a practical DBS service

The success of DBS is intimately linked with the technical standards employed. It not only determines the resolution and fidelity; it also influences the satellite power and the system performance under adverse conditions.

At the time of these planning conferences work had to be based on the existing terrestrial standards of PAL/ SECAM and NTSC. There were, as yet, no firm plans for DBS services but, in order to provide for the future

186

development, it was agreed that all countries would only
introduce new standards if they caused no increase in
interference to the other services.

The only features considered here in a quantitive
manner are the signal to noise ratio of the received
signal and its freedom from co-channel and adjacent
channel interference.

The main difference between composite and component
systems arises from the presence of a high frequency
carrier in composite systems. Although composite systems
make efficient use of the available bandwidth, experiment
has shown that, for a given satellite power, a component
type of system has an advantage (see Chapter 5).

Furthermore a component system is likely to benefit
most from the technique of threshold extension. As
shown in Fig.4.14 this advantage is most significant
when considering the weak signals associated with
reception outside the normal service area.

Experiment has also shown that there are no significant
penalties in the interference performance. Furthermore
if scrambling is employed to implement a "conditional
access" system this may result in a further reduction in
interference. This could be important in some parts of
Region 1 where, due to overcrowding, the interference
specification has not been fully met.

4.9 OPTIMUM TRANSPONDER POWER IN THE U.K.

However the cost of a satellite is very dependent on the
power of the transponder and the choice of satellite
power is inevitably a compromise. Anyone motivated
solely by the need for profit will choose a low power

compromise. Similarly those that seek to provide a
public service, and leave no member of society deprived,
will choose a high power compromise. Hopefully, all
practical examples will result in a compromise which
lies somewhere between these two extremes.

Estimating on the basis described elsewhere in this book
it is clear that if a power of, say, 30 watts is used
the result will not be a satisfactory service by any
standard. Equally, many will see a good justification
in providing the full power offered by the WARC and
reducing the burden on the individual receiver to a
minimum.

If circumstances are such that a compromise must be made
the lowest reasonable figure for the UK seems to be about
100 watts radiated power but this will result in some
deprivation for the less fortunate viewers.

Further discussion on optimum transponder power is to be
found in Chapter 5.

Chapter 4

REFERENCES

No.	Subject	Book Reference
1	The C-MAC/packet system for direct satellite television H. Mertens & D. Wood Published by Technical Centre E.B.U. Brussels	4.8.1
2	Satellite Broadcasting: system concepts and the television transmission standards for Europe G.J.Phillips & P.Shelswell Int.Journal of Satellite Communcations	4.5
3	International Telecommuni- cations Union 1977 World Broadcasting Satellite Administrative Radio Conference Final Acts, Geneva, 1977	
4	Direct Broadcasting from Satellites G.J. Phillps I.E.E.Proc. Vol.129, Pt.A, No.7, Sept 1982	4.1.4
5	High-definition television for satellite broadcasting E.B.U. Review - Technical No.170. August 1978	4.4.5

CHAPTER 5

The Domestic Installation

5.1 A REVIEW OF THE OPTIONS

In the days of the WARC planning conference there were
many unsolved problems associated with DBS. As yet
receiver technology had not provided an answer to the
mass production of 12 GHz devices. The Gallium Arsenide
technology that was to provide the solution to high
frequency amplification problems, and make it possible
to construct low noise amplifiers, was still in its
infancy.

There were those who had a naively simple view of the
future. They foresaw a local oscillator and frequency
changer followed by an RF conversion from Frequency
Modulation to Amplitude Modulation. The baseband
standards were such that this signal could then be fed
directly to the conventional terrestial UHF receiver.
As we shall see later converters were made of this type
but they did not perform very well.

In the end new integrated circuit technology came to the
rescue and made it possible to make high performance
devices at a price that the domestic market could afford.
The concept of launching the DBS service using a
converter attached to an existing terrestial receiver

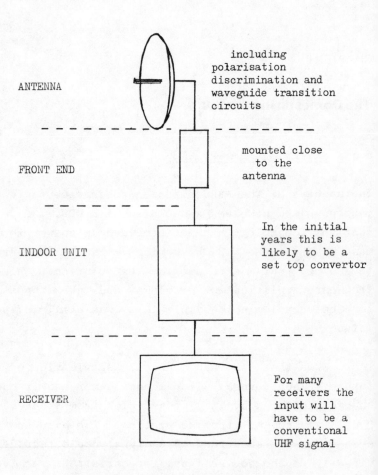

ANTENNA   including
polarisation
discrimination and
waveguide transition
circuits

FRONT END   mounted close
to the
antenna

INDOOR UNIT   In the initial
years this is
likely to be a
set top convertor

RECEIVER   For many
receivers the
input will
have to be a
conventional
UHF signal

Fig 5.1. Component parts of a DBS receiver

remains with us but it is now a relatively sophisticated device.

This converter consists of three units (Fig.5.1)
1. The antenna
2. The outdoor unit
3. The indoor unit

It is first necessary to set the scene before embarking on the detailed description of these three units.

5.1.1 New Services

Satellite broadcasting has the major advantage that all of the country can be served by one transmitter. However cable systems exist and provide a unique service offering a wide choice of programmes. It may be that the cable companies will wish to carry the DBS services (or may be required by law to do so - the so-called "must carry" rule). In addition there will be those that live on the wrong side of a block of flats and cannot "see" the satellite. In these cases the economic pressures on the satellite installation are different. The financial burden of buying a larger antenna for cable systems is of little consequence.

In countries where no full power DBS service is provided these will be those who offer an economic service for community reception using a low power satellite. The WARC prescribed a power flux density of -103 dBW for domestic reception and -111 dBW for these community reception systems (Regions I and III). This suggests a community antenna of about 2 metres diameter and the 12 GHz plan was checked for this.

However it should be noted that there are additional, separate frequency bands allocated for community reception. Furthermore distribution to the cable head receiver is usually regarded as a "point to point" transmission and will be subject to different regulations to the DBS services (see Chapter 6 on legal issues)

The presence of cable services might be considered to be of particular importance when considering the technical standards. Cable systems are subject to different technical restraints. In general terms cable systems do not suffer from difficulties arising from low signal-to-noise ratios but they do sometimes have severe bandwidth limitations. They also suffer from reflections due to cable irregularities or poor terminations.

Cable systems are of different types. The small distribution systems in a block of flats may well distribute an FM signal on the first IF (Intermediate Frequency).

However the larger system may well carry a variety of services and bandwidth restrictions may force those concerned to distribute using a Vestigial Sideband Amplitude Modulation System (VSB). Unfortunately this system can introduce low frequency distortion which might make it difficult for some cable systems to accept a scrambled signal. The "Switched Star" type of cable system does not have to carry a scrambled signal and does not suffer these problems.

Thus there may be many different types of receiver and the technical standards adopted may have to be designed to meet a broad range of requirements. The receiver described in the following pages is primarily intended for domestic use and employs the C-MAC coding system. However receivers on many other standards will employ similar techniques and much of the comment will be appropriate whatever standard is used.

5.1.2. New Standards

It is not necessary to include detailed design information on the terrestrial standards because this already exists in many textbooks. Unfortunately the new DBS standards are still subject to change but the following observations on the receiver specification are appropriate.

Vision

The receiver employing new standards will make great demands on the integrated circuit industry. It will include one, or more, microprocessor and contain a number of integrated circuits specifically designed for this receiver. To a great extent the technical complication within the receiver is obtained without an increase in price providing the market is large enough to allow recovery of the development costs of these integrated circuits. This new technology has made it possible for engineers to optimise the coding of a television picture and to provide a range of new facilities. They have also made provision for a better "user interface" which will help the viewer operate this new receiver and find the right programme in spite of the variety of choice and the technical complications involved.

The members of the European Broadcasting Union have
devised such a system (described in earlier chapters as
"C-MAC"). The overall structure of such a receiver is
given in Fig.5.12. Further information about the signal
processing that takes place in the blocks within this
diagram is to be found in the appendices.

In the meanwhile many countries continue to work on High
Definition Television and, if this is successful, some
DBS allocations will be used for this purpose. This
will require a major change to all aspects of the
receiver design including the television display itself.

The current trend is to divide the problem into segments.
We may have up to four segments:

        1. The Studio Programme source
        2. The Distribution System (transmission)
        3. The Broadcast System
        4. The Receiver Display

All of these segments may employ different High
Definition Standards but they will be chosen so as to
make the interface (or conversion) at each junction a
relatively simple one.

The Broadcast Standard will have to fit within a limited
bandwidth and the receiver will have to contain the
signal processing that will make the most of this
limited information. Such a development is closely
linked with the introduction of digital processing in
the receiver and the availability of low cost picture
and sound storage.

Sound

In the days of the WARC terrestrial standards were
assumed for both the vision and sound signals. The

sound would therefore be carried on a subcarrier above
the video signal.   In terrestrial standards this is
typically at 5.5, 6.0 or 6.5 MHz.

As already described both the satellite vision and
sound signals have to be frequency modulated on the
12 GHz carrier.

In the intervening years since the conference there have
been several major developments in the domestic market-
place.   There has been the growth of the domestic
recorded (VCR) and there has been the successful
introduction of digital sound recordings.   We have also
seen the introduction of stereo sound to accompany the
television signal.   Any new service cannot ignore these
developments and all the proposals for DBS make some
provision for stereo sound.   Digital coding is a
favourite form of modulation which has only minor cost
consequentials in this era of integrated circuits.   Thus
we are likely to see a situation where all DBS standards,
be they existing terrestrial standards, or new improved
performance colour component systems (or high definition
television), they will all employ digital sound coding.
Proposals already exist in Japan, North America and
Europe for such standards.

This will probably mean that the converter, or adaptor,
for existing television receivers will have a "Hi Fi"
stereo output in addition to the conventional PAL, SECAM
or NTSC UHF modulated output with the conventional sound
carrier.

This concept is compatible with industry's tendency to
provide domestic equipment in the form of modules, or
units, which may be added together to meet customer
requirements.   It is expected that these converters will

also include a "Cenelec" 'peritelevision' connector
(for both the sound and vision signals) which will
enable a variety of such units to be connected together.

Without doubt this receiver of the future will also have
a standard remote control system and hopefully this will
be compatible with the VCR. In addition there will be a
market demand for a converter which provides an output
for the VCR so that the viewer may record one service
while watching another.

Before discussing the detailed structure of this
domestic receiving terminal, it is worth while briefly
recalling, once again, the fundamental constraints
imposed on the DBS system by the WARC plans and to
consider their impact on the quality of television that
can be produced and hence the design of the receiver.

First, the choice of the geostationary orbit allows the
receiving aerial to be of high gain yet stationary,
and it fixes the clear-weather path loss or 'spreading
loss'.

The choice of a frequency around 12 GHz avoids the worst
effects of rainfall attenuation while minimising the
physical size of the transmitting and receiving aerials
for a given power gain and bandwidth. The transmitting
aerial size and gain are dictated by the beamwidth
necessary to concentrate the power over the area on
earth that it is to serve; the receiving aerial size is,
on the other hand, limited to around 0.9 m by considerations of windage and pointing prevision, leading to a
maximum gain of about 38 dB, yet affording adequate
discrimination against interference from adjacent
satellite positions 6 degrees away.

The choice of 12 GHz was also a challenge to the electronics industry to produce domestic receivers with the highest performance the reliability at the lowest cost in manufacture and maintenance; what they have achieved has certainly surpassed the dreams of the early planners of 1971.

Taking all these factors together, we are able to calculate how much power must be radiated by the satellite transmitter so that the carrier power in the domestic receiving terminal is much greater than the receiver noise level and a good quality of service is enjoyed for most of the time.

5.2 PERFORMANCE CALCULATIONS

5.2.1. Noise Factor, Noise Figure, Noise Temperature

Perhaps the most important feature of the performance of the receiver is the level of the inevitable, random noise power that arises in the circuitry of the 'front-end', putting a lower limit to the level of receivable signals.

One way of defining receiver noise is to compare it with a known source of noise, such as a hot resistor, which generates a random noise power that is proportional to its absolute temperature.

Thus, if we imagine that the receiver is fed from a resistor of the same impedance as the aerial but at a temperature of 290 K (a conventional standard for this purpose), then the Noise Factor, Noise Figure or NF of the receiver is given by:

$$N = \frac{\text{total noise power}}{\text{noise power from resistor alone}}$$

and usually expressed in decibels.

Clearly, a perfect receiver has a noise figure of N=1
(or O dB) and the contribution of the receiver alone is
N - 1.

Another way of expressing receiver noise is to attribute
to it a temperature, as if it were coming from a built-
in hot resistor at its input. Thus, the noise temperature of a receiver is related to its noise figure by

$$T_r = (N - 1) 290 K$$

so that a noise figure of, say, 3 dB (N = 2) corresponds
to a receiver noise temperature of 290 K.

5.2.2 Figure of Merit (for a receiving terminal)

But of course there will be other sources of noise in
the system. Some of these we must discount for the
present if they are actually transmitted by the up-link.
But some noise will accompany the reception of the
wanted signal - not because the aerial is a hot, lossy
resistor (ohmic losses in the aerial are usually very
low) - but because the aerial is 'looking at' a
transmitter with a surrounding background of hot earth,
hot air or hot space, as was explained in Chapter 3.

For example, the receiver in the broadcast satellite is
'looking at' the Earth around the earth-station transmitter so that the lowest noise temperature is limited
to that of the earth (around 290 K) whereas the domestic
down-link receiver has an effectively cooler source
because its aerial looks past the satellite transmitter
into space at a temperature of around 20 K (provided
that the air is clear and not overheated by a rain-
storm). In this connection it is interesting to note
that the effects of rain-attenuation on the signal are
accompanied by a rise in the noise temperature of the

sky (Fig.3.2) so that the carrier-to-noise ratio will deteriorate to a greater extent than the attenuation alone. In fact, if the noise figure of the receiver is less than about 2.6 dB, then the rise in noise has a greater effect than the loss of signal.

To summarise the situation, the effective temperature of a receiving terminal, T, is the sum of all the temperatures of the relevant sources of noise, suitably enhanced or diluted according to the gains or losses between them - the source, the aerial, the feeder and the front-end of the receiver. But the ability of the receiver to raise the wanted signal carrier power to an acceptable level is determined by the power gain of the receiving aerial (less losses), G.

So a useful 'Figure of Merit' for a receiving terminal may be written simply: G / T.

The reader should take care not to equate G/T with the actual carrier-to-noise ratio when comparing receiving terminals of different bandwidths (e.g. different transmission systems) because the noise power will be proportional to the receiver bandwidth whereas the signal carrier power will not.

Finally, it should be remembered that G/T is dependent on the weather because the effective temperature, T, must include not only the temperature of the losses of the aerial but also the temperature of the sky that it looks through. The Table in Fig.5.2 shows how G/T depends on the noise figure of the receiver both in fine weather and in conditions of heavy rain.

D B S Receiver		System, Temp. and G/T			
		Clear Air		Heavy Rain	
Noise Figure dB	Noise Temp K	Temp K	G/T dB	Temp K	G/T dB
8	1540	1595	5.5	1705	5.2
6	865	920	7.9	1030	7.4
5	630	685	9.1	795	8.5
4	440	495	10.6	605	9.7
3.5	360	415	11.3	525	10.3
3	290	345	12.1	455	10.9
2.5	225	280	13.0	390	11.6
2	170	225	14.0	335	12.2
1.5	120	175	15.0	285	13.0

Fig. 5.2. System G/T : dependence on
Noise Figure and weather

Fig. 5.2. makes the following assumptions:
- 38.5 dB Antenna gain (55% efficient,
0.9 metre dish)
- 0.5 dB Antenna Coupling loss
- 0.5 dB Antenna pointing loss
- 150 K Antenna noise temperature

5.2.3 Carrier to noise calculations

As we shall see later the Carrier to Noise ratio (C/N)
is a measure of the performance of the front end of the
receiving installation. This is the ratio of the signal
power collected by the antenna and the noise power as
represented by the system temperature, that is:

$$\frac{C}{N} = \frac{PFD \times Effective\ Antenna\ Area}{(k \times B \times T)}$$

In this equation PFD is the power flux density arriving at the antenna (W/m^2) and B is the receiver bandwidth (Hz).

As the theoretical antenna gain (G) is

$$\frac{4 \pi \times \text{Effective Antenna Area}}{\lambda^2}$$

this equation may be rewritten as:

$$\frac{C}{N} = PFD \cdot \frac{c^2}{4\pi f^2} \cdot \frac{1}{kB} \cdot \frac{G}{T}$$

where the constants are

c, the velocity of an e.m wave in free space, which is 2.9988×10^8 metres/second

k, Boltzmann's constant, relating molecular kinetic energy to absolute temperature which is equal to 1.3803×10^{-23} WS/K

f, the frequency of the wanted signal (Hz) and λ its wavelength

Expressing this in decibel units we have:

$$\frac{C}{N} = PFD \ (dB \ W/m^2)$$
$$+ \frac{G}{T} \ (dB \ 1/K)$$
$$+ 111 \ db$$

where the receiver is operating at 12 GHz and has a bandwidth of 27 MHz.

This calculation assumes that the figure for power flux density makes appropriate allowances for attenuation in the atmosphere.

It makes a small allowance for noise introduced in the up-link (about 1/2 dB).

5.2.4. Relationship between Carrier-to-Noise Ratio, Noise Weighting and Picture Quality

We need to know the lowest ratio of carrier power to noise power that will give a picture of adequate quality for broadcast reception. There are three steps in the conversion from C/N to picture quality; these may be described as follows.

(1) C/N to S/N_{uw}. Here the C/N is simply a radio-frequency power ratio but the S/N_{uw} refers to a video frequenty measurement of peak-to-peak luminance/unweighted r.m.s. noise. This part of the conversion is entirely objective and may be calculated from a knowledge of the modulation system and the bandwidths, as is shown by the following example for frequency-modulated, standard-I signals.

$$\frac{S/N_{uw}}{C/N} = \frac{3}{2}\left(\frac{f}{f_v}\right)^2 \frac{B}{f_v}$$

where   f = frequency deviation
            f_v = video bandwidth
            B = r.f. bandwidth

(This system has a synchronising waveform whose amplitude is 30% of the transmitted signal)

Putting in the values for the PAL system, the above ratio comes to about 16.5 dB but an additional factor of about 2 dB must be added to include the

effect of pre-emphasis. (Pre-emphasis refers to
the accentuation of higher-frequency video
components before transmission - according to an
agreed characteristic recommended by the CCIR - so
that, when the signal is corrected by the corresponding, inverse or de-emphasis, network in the receiver
some of the high-frequency noise components are
reduced in power, thus improving the S/N ratio).

A similar calculation may be made for the
chrominance signal for it is quite possible that,
with certain systems and certain highly saturated
colours, the chrominance signal-to-noise ratio may
determine the overall picture quality.

As might be expected, two different systems of
transmitting luminance and chrominance would show
different ratios of S/N to C/N; the results, after
correction for pre-emphasis, give a figure of
18.5 dB for PAL (1) and about 16 dB for C-MAC.

(2) Unweighted/Weighted Signal-to-noise ratio
S/N_{uw} / S/N_{w}

Because of the differences between the objective
magnitude and the subjective effect of noises of
different spectra accompanying the video signal, a
systematic 'weighting' has been developed so that,
knowing the spectrum of the noise (determined by the
modulation system and the use of pre-emphasis), a
hypothetically-weighted S/N may be established that
may be directly related to the subjective impairment
of the picture. Alternatively, a 'unified weighting
network' may be constructed, through which objective
measurements may be made that give a result
equivalent to the subjective effect produced.

So we may establish 'weighting factors', depending on the system under consideration; for PAL(I) the factor is 11 dB and for C-MAC the factor is 13 dB, showing that the systems have a comparable subjective performance for the same value of S/N_W.

(3) Weighted S/N / Picture Quality

If the 'weighting' factors are correct, then there should be a fixed relationship between the S/N_W and the picture quality - this is the 'crock of gold' that is sought by all those who engineer and compare improvements in television systems. But the end of the rainbow is still rather elusive, even after years of exhaustive (and exhausting) subjective tests in which the subjects were asked to 'grade' a particular picture according to a list of numbered verbal descriptions.

Fig.5.3 shows the approximate connection between C/N, S/N_W and picture quality; the results apply to the C-MAC system chosen for DBS, though the PAL system is only about ¼ grade lower in performance.

It is possible to improve the quality of the received picture at low values of C/N through the use of 'threshold-extension' techniques that increase the effective C/N by reducing the noise bandwidth in the receiver, leading to up to one grade improvement at C/N values of around 6 dB. This technique may be applied very effectively to the MAC signal.

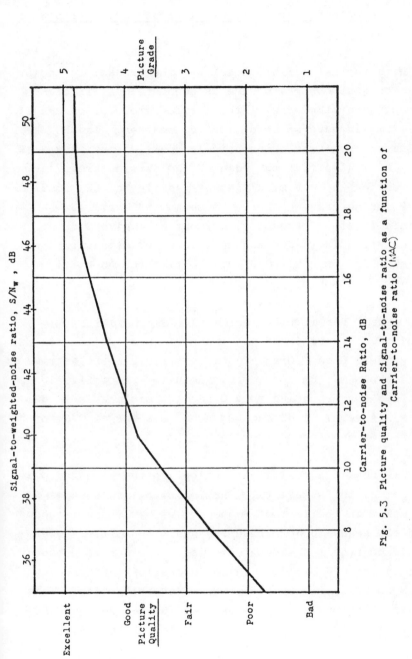

Fig. 5.3 Picture quality and Signal-to-noise ratio as a function of
Carrier-to-noise ratio (MAC)

5.2.5 Reception standards; satellite transmitter power; dish size

As with terrestrial broadcasting, the transmitter power must be just enough to give an adequate quality of reception at the edge of the service area for enough of the time in adverse conditions of weather. In DBS the required margin is comparatively small, comprising 3 dB for the fall-off at the edge of the service area, say 0.5 dB for antenna misalignment and (in the UK) about 2 dB for adverse weather. As shown in Chapter 3, a rainfall rate of around 20 mm/hour will give 2 dB attenuation in the UK and this level of attenuation will be exceeded for only 0.3% of the time (or about 1% of the wettest month).

If, except for a small period of heavy rainfall, the quality of reception at the edge of the service area may be graded 'Good' (Grade 4) or better, then the service as a whole may be considered adequate. From Fig.5.3 the C/N corresponding to a Grade 4.3 picture is 14 dB so a figure of 14 dB may be taken as a target C/N for reception in these circumstances.

We are now in a position to calculate the required power output of the satellite. The following table shows an example, in the form of a budget, of how the noise power and the signal power arise and what transmitter power would be required at the satellite in order to produce the target C/N of 14 dB in the domestic receiver with an 8 dB noise figure and a 0.9 m receiving dish. This table therefore ambodies the thinking behind the WARC 1977 Plan, as it applied to reception in the UK.

U.K Down-Link Power Budget

Noise
Receiver temperature (8dB NF)	1540		
Aerial . . heavy rain	110		
0.5 dB loss	35		
Galactic	20		
(see 5.2.2)	1705 K	32.3	dBK
Receiver bandwidth (27 MHz)		74.3	dBHz
Thermal noise (Boltzmann's constant)		-228.6	dB W/Hz
Noise Power in Receiver		-122	dBW

Carrier-to-noise ratio for DBS reception
(exceeded for 99% worst months)
(see 5.2.3)   14   dB

Carrier power in receiver
(at beam edge in heavy rain)   -108   dBW

Domestic aerial area (0.9m dia.			
55% efficient. 38.5 dB gain)	-4.5		
coupling loss	0.5		
		-5	dBm^2
Power Flux Density at edge of beam		-103	dBW/m^2

Beam Size for satellite to cover UK
1.8 x 0.7 degrees

Satellite aerial Gain for this beam	43		dB
(see 5.4.1)			
loss at beam edge	-3		dB
atmospheric loss	-2		dB
(see Ch.3)			
spreading loss	-163		dBm^2
(at 40Mm range)			
	-125	-125	dBm^2
Transmitted power from satellite		22	dBW
			(160W)
EIRP (power x axial gain)		65	dBW

208

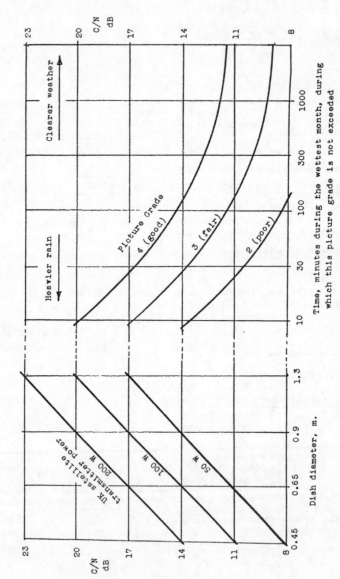

Fig. 5.4 Effect of changing transmitter power and
receiving dish size on the C/N and picture
grade received in London with a receiver
having a noise figure of 5 dB

However, as we know now, the receiver noise figure is likely to be much lower than 8 dB so that it is appropriate to ask whether we may make a 'saving' in some other direction - such as the power of the satellite transmitter or the diameter of the receiving dish. Fig 5.4 shows how the two latter variables affect picture quality as received in London with a receiver having a noise figure of 5 dB in conditions ranging from 'clear weather' to heavy rain.

5.3 <u>RECEIVER CONFIGURATION</u>

As described previously the receiver (or converter) is considered to be in three parts. As illustrated in Fig.5.1 this consists of:

    1. The antenna

    2. The outside unit (presumably mounted on
                       the antenna)

    3. The indoor receiver (or converter)

After the initial collection, frequency conversion and amplification of the signals coming from one particular satellite orbital position, the whole 800 MHz band (or at least half of it) is cabled at i.f. from the outdoor unit to an indoor unit for filtering, demodulation, distribution and display. There will be one basic limitation to the scope of the choice of signals to the indoor unit (or for FM distribution) - only those signals

transmitted on one of the two polarisations will be
available at any one time (about half of them), the other
half will require a duplicate receiver, fed either from a
second antenna or from a dual-polarisation antenna. With
this exception, the whole band may be available for
subsequent selection in the indoor unit.

Even though all of the 40 channels may thus be available,
the primary aim of the planning is to give priority to
the smaller number of channels assigned to that particular area, together with an optional selection of those
channels assigned to an adjoining area. If, for example,
you live in West Germany, you will be mainly interested
in Channels 2, 6, 10, 14 and 18 but, if you are close to
the Austrian border, you would also expect to be able to
receive their programmes on Channels 4, 8, 12, 16 and 20.
They are transmitted with the same polarisation but are
separated by two channels from your own which is enough
separation to avoid adjacent-channel interference. However, if you are prepared to spend more money on the
installation, you may be able to switch the polarisation
of the antenna and also receive the French channels 1, 5,
9, 13 and 17, without having to alter the antenna pointing.
There are thus several different options open to the
designer of the domestic in-house units, depending on
geography, cost and individual preference.

A further important consideration is how to provide for
a new generation of television and sound receivers
without rendering existing receivers in any way redundant.
This means that the initial requirement for satellite
broadcast reception must be an adaptor or converter which
is to be inserted between the existing receiver and the
aerial downleads, enabling it to display either terrestial or satellite programmes.

As we shall see later the indoor converter is a complicated device and, if the standards employed for DBS are
different from the terrestrial standard, the ultimate
receiver which can display both satellite and terrestrial
television will also be very complicated. The three
basic components will now be considered in detail.

5.3.1. The Antenna

Looking again at the basis of the WARC down-link power
budget the effective area of the receiving aerial
aperture must be about -5 dBmetre^2 to give satisfactory
reception when the power flux density is -103 dBW/m^2.
Probably the simplest aerial that satisfies this requirement (and meets the general specification laid down by
the planners) is a shallow paraboloid or dish 0.9 metre
in diameter (3 ft) with a small pick-up aerial or waveguide horn mounted at the focus. The pick-up horn or
feed cannot be highly directional and it collects the
power from as much of the dish as possible without
'looking' over the edge of the dish where there may be
a source of interference. In practice the effective
area of the dish will be less than the actual area; an
efficiency of 55% to 65% is common. Figs.5.5 and 5.6
show four different ways of arranging the pick-up from
the dish.

The pick-up antenna must be polarised so as to have
greatest sensitivity to circularly-polarised waves of
one direction, rejecting those of the other direction.
The circular polariser may take a wide variety of forms,
from a simple arrangement of crossed dipoles and delay
lines to a printed-circuit spiral or a more complicated
arrangement of waveguide junctions, some of which allow
both polarisations to be extracted from the same antenna.

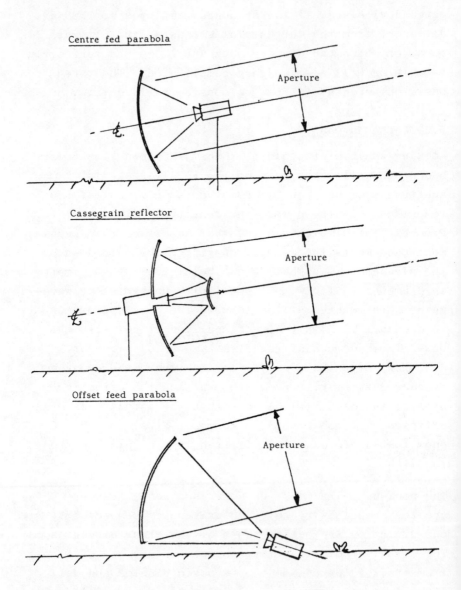

Fig. 5.5 Antenna Configuration

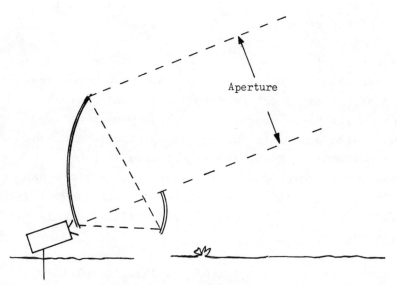

Aperture

<u>Fig. 5.6 Offset Gregorian Antenna</u>

Efficiency of up to 68% has been quoted for this
antenna.

Whatever form of polariser is used, its performance must
be within the templets specified by the planners in order
to avoid interference, and care must be taken in the
design of the structure so as to avoid losses.

The accuracy of the surface of the dish is important but
the departure of the surface from that of a true paraboloid of $\pm$ 1 mm is unlikely to have any significant
effect on its performance. The popular image of the
domestic dish antenna is that of a dustbin lid and many
people contemplate using aerials of equally poor
construction. However, it is the off-axis performance
that determines whether a satellite-receiving antenna will
be satisfactory in the presence of signals from adjacent
satellites.

If the dish diameter is reduced the signal amplitude
will fall and the ability to reject unwanted signals
from adjacent satellites will decrease. For those
countries that need only limited selectivity the antenna
can be reduced to 0.6 metre diameter. There has been
some consideration of dishes 0.45 metres diameter but
these may give a very poor signal-to-noise ratio and
little rejection of interfering DBS signals.

It may be a habit, formed during the regime of terrestial
broadcasting, that if you live in a large town you can
'get away with' a poor antenna. Unfortunately, the same
does not apply to satellite broadcast reception for
there is little difference between reception in town and
country. The same antenna performance is required for
good reception in the cities as outside in the country,
permitting no concession to ease the aesthetic problems
presented by a congestion of three-foot diameter dishes
in built-up areas.

There is therefore a strong incentive for designers of domestic antennas for satellite-broadcast reception to produce a design that complies with the basic requirements of the overall planning and is, at the same time, more acceptable in appearance and convenience than the basic three-foot dish. Ways of reaching this objective are discussed later.

Required antenna pointing accuracy

Using simple antenna theory, the gain of an antenna is related to the number of square wavelengths contained within its relative entrance pupil by the expression:

$$G = 4\pi A / \lambda^2$$

$$= \left(\frac{\pi D}{\lambda}\right)^2 \cdot \frac{P}{100}$$

Where: G is the axial gain over an isotropic antenna

A is the effective area

D is the diameter (metres)

is the wavelength

P is the percentage aperture efficiency

Using the same estimates of practical performance as
were proposed and adopted by WARC, 2, the half power
beamwidth of an antenna is 1.78/D degrees at 12 GHz.
It follows that a 0.9 metre dish has about a 2 degree
beamwidth. A 0.5 degree pointing error should not
introduce more than a 1 dB loss. A pointing tolerance
of 0.5 degree would therefore be considered tolerable
provided that other losses had already been minimised.
Such a stability is fairly easy to achieve if the dish
is mounted on a rigid support and fixed to a wall,
chimney block or the ground. On the other hand, the top
of a 30-foot mast would be unlikely to provide the
stability necessary to keep the satellite signal within
1 dB of the maximum.

Practical Antenna Alignment

If one surveys the installation of most television
aerials it is clear that a more secure fixing and a more
accurate alignment will be necessary for the DBS antenna.
It has been suggested that the receive antenna pointing
accuracy should be $\pm$ 0.5 degrees and this will not be
easy to achieve. Vaguely searching the sky for a signal
is most likely to capture the wrong satellite. However
there are several ways that this problem can be overcome.

1. The antenna can be fitted with reference lines
and these can be set at the required elevation angle
using an inclinometer. This leaves only one degree of
freedom (azimuth) to search the sky (see Fig.5.6). If
the search path crosses the geostationary orbit at a
large enough engle it should be possible to find the
right satellite. It should be noted that the installation angles vary throughout the country, as shown in
Fig.5.8.

Fig.5.7 Path of the sun relative to the

geostationary orbit

Summer

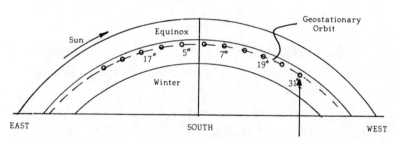

EAST

SOUTH

WEST

Position of the sun
on March 2nd and
October 13th at about
2.30 pm as seen from
London

see Appendix 8

218

Engineering Information

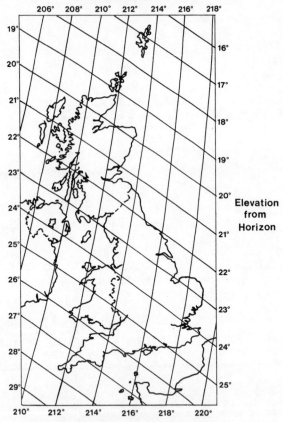

**RECEIVING AERIAL POINTING CHART FOR
UNITED KINGDOM BROADCAST SATELLITE
(Orbital Position 31°W)**

**Elevation
from
Horizon**

Azimuth Bearing East of True North

The chart shows the required azimuth bearing (from true north)
and the elevation angle (from horizontal) for the boresight direction
of a receiving aerial according to its location in the United Kingdom.

Information Sheet 5003 (2) 8305

Engineering Information Department, BBC, Broadcasting House, London, W1A 1AA   01-580 4468 ext 2921
Data prepared by BBC Engineering Research Department

Fig 5.8

2. The antenna can be fitted with a horizontal axis such that the alignment can be achieved using a plumbline and spirit level. This presents a production problem because the antenna design will differ depending on the location.

3. We know that the antenna can be aligned by pointing at the sun on March the 2nd or October the 13th at 2.30 p.m. (see Fig.5.6 and appendix).

It is also possible to calculate the position of the satellite relative to the sun at any more convenient time of day (see References).

In all of these cases a compass will help ensure there are no gross errors but, inevitably, some trial and error adjustment will be necessary. In order to make this possible some manufacturers are considering including a wire in the downlead which will carry the AGC (Automatic Gain Control) signal. A portable meter can then be used for the final alignment.

The Stepped Reflector

One way in which the bulky, paraboloidal dish may be transformed into a less obstrusive structure is to use a stepped reflector, as shown in Fig. 5.9. The object of this design is to construct a reflecting surface out of several coaxial paraboloids, axially separated by multiples of one wavelength. In this way, the phase front of the wave may be planar, although broken up into separate rings. The performance of such a reflector is similar to that of a continuous paraboloid, except that there is a small additional loss, owing to the waste of the energy falling on the outward-facing areas between the steps. There is also a limitation in the bandwidth

of the reflector. Nevertheless, the stepped reflector
design offers a much 'flatter' antenna that may be less
prone to catching the eye (or the wind), and therefore
more attractive environmentally. An offset version of
the stepped reflector may be designed, so that the
axis of maximum reception is at an angle of, say, 80
degrees to the plane of the antenna, allowing a greater
latitude in mounting the antenna close to a fixed structure
yet at the correct angle.

Antennas with stepped reflectors are sometimes referred
to as Fresnel antennas because of the similarity of the
geometry of the stepped areas to those of a Fresnel lens
or a zone-plate where focusing is obtained by controlled
diffraction.

Toroid Reflectors

Although theoretically it is necessary to have a parabolic reflector it is possible to make approximations
if the aperture is small. Approximate reflectors are
inevitably less efficient and suffer from aberrations.
However, it is frequently convenient to make the simple
dish reflector spherical in shape.

These aberrations may be reduced by using multiple feed
points, each covering a small aperture. However, because
this is relatively expensive, it is unlikely to be used
in the domestic market.

Furthermore, if it is required to slew the antenna,
there may be an advantage in using a cylindrical component. Thus toroidal antennae have appeared on the
market where one axis is parabolic and the other is
cylindrical. Such an antenna could, for example, slew

Fig 5.9 Stepped Reflector
courtesy BBC

Fig 5.10 Phased array
courtesy BBC

to cover geostationary satellites many degrees apart
(from 19 degrees to 31 West for example).

This type of antenna can,in principle,be constructed in
the stepped form so as to make a substantially flat aerial.
Also,the ability to slew the aerial would mean that it
could be mounted flat on a convenient wall even though
the wall was not exactly at the correct angle. It would
then be convenient to mount offset feed points on the
ground and the receiver could switch between them when a
change in orbital position is required. Clearly such an
antenna would be much more expensive than the simple dish.

The Phased Array

The aim of the two foregoing aerial designs has been
to use a reflecting surface of the correct size to
condense the power of a plane wave on to a single pick-
up point or feed at the focus of the reflector. Providing
that the phase-front of the incoming wave is normal to
the axis of the reflector, then all parts of the wave
will arrive in the same phase, reinforcing the field at
the pick-up aerial. At any other angle of arrival,
there will be mis-phasing and destructive interference,
giving the aerial its directive properties.

However, instead of condensing the incident power at a
focus, it is possible to collect it in an array of
separate, simple aerial elements distributed over a plane
surface, adding the power in a network of conductors
behind the surface. By making the array from a sandwich
of copper and dielectric, like a printed circuit, the
aerial elements may take the form of slots or holes in
the outer surface, connected by a series of printed
transmission lines to the common output.

A fundamental difficulty in the making of a planar array
of printed elements is the choice of 'substrate' - the
material that keeps the metal conductors apart and allows
the power to flow with the least loss. Unfortunately,
suitable materials are too costly for a domestic design
and if no breakthrough in solid dielectrics occurs then a
metal structure with an air dielectric would seem to be a
possible solution.

A second problem in the construction of a planar array of
discrete elements (slots or apertures) is that, in the
case of satellite broadcast reception, each element must
be circularly-polarised in one direction or the other.
The construction of circularly-polarised apertures and
their application to a planar array is not easy. The
most likely device for this purpose seems to be the
addition of a parasitic element to each slot having the
appropriate spacing and reactance so that it enhances the
one polarisation while suppressing the other. Alternatively a secondary array or screen can be put over the
planar antenna to suppress the unwanted polarisation.
The form that this device will take is a matter of
continuing experiment and development, until a reliable,
cheap method is found.

Thirdly, a regularly-periodic array of elements, all
making the same power contribution to the output, is not
necessarily the ideal. The planning specification laid
down by the WARC for the domestic receiving aerial would
not be met by such an array unless either the spatial
distribution or the power contribution of the elements
is disturbed so that the effective aperture was 'tapered
off' at the edges. This 'tapered aperture' may tame the
side-lobes of the radiation pattern of the aerial but it
also results in some loss of efficiency.
A possible way of utilising the propensity of a discrete
array to produce large side-lobes is to arrange that the

nulls between the side-lobes fall neatly on the direction of the adjacent satellite; these are (almost invariably) at 6 degree intervals. To make this occur with a uniformly-illuminated aperture at 12 GHz, the width of the aperture (East-West) must be a multiple of 10 or 25 cm. Some flexibility in the shape of the aperture is then possible, the vertical dimension (North-South) being adjusted to make up the required area. It should be emphasised that there may be other sources of interference than those at ± 6 degrees, for which the full WARC specification may be required.

Lastly, there is the possibility of an 'active' array of elements in which each of the individual elements is closely-coupled to a r.f. amplifier. For this purpose low-noise Gallium Arsenide amplifying devices may become so cheap and reliable that the cost of an array of a few hundred of them would be offset by the opportunity to follow the amplifiers with relatively lossy transmission lines, but this seems unlikely.

5.3.2 Front-end Electronics

The front end receives a signal at 12 GHz and amplifies it. It then passes to a mixer to be changed in frequency to that used by the first IF amplifier before passing it to the downlead.

The polarisation discrimination has to be built in the antenna feed point (or more accurately the collection point) and there follows the transition coupling from the waveguide to the electronic circuits. The front end amplifier will determine the signal-to-noise ratio and provide any filtering that is necessary.

226

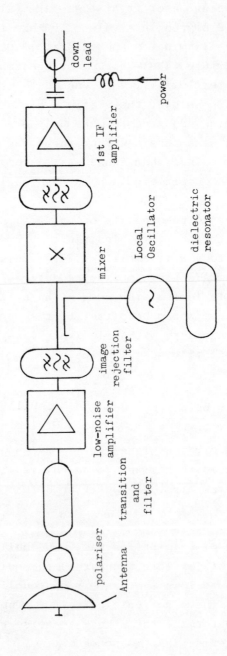

Fig. 5.11 Receiver 'front end' schematic

The WARC conference assumed a NF of 8 dB. It has been shown that, with a normal receiver, this will result in a figure of merit (G/T) of 6 dB for a 0.9 metre antenna, 55% efficient. There has been a steady improvement in these figures over the years.

Initially work was concentrated on a planar form of construction with discrete components. It was difficult to obtain better than 4 dB NF using these techniques.

Such a front end would have a two-stage Gallium Arsenide FET (Field Effect Transistor) amplifier following the impedence matching circuits.

There would then be a mixer and local oscillator using a GaAs Schottkey barrier diode and a dielectric resonator using a Gunn diode.

Finally there would be a wide band IF amplifier.

Using monolithic construction and the latest GaAs
techniques it is now possible to get an excellent Noise
Figure of 2.6 dB. Significant improvement beyond this
figure may be difficult.

The necessary rejection of image frequencies is not
difficult to obtain and the adjacent channel rejection
is provided by the indoor receiver.

5.3.3. The down lead

The power required for the front end will have to be
sent up the 'down lead' or an associated cable. This is
a minor problem but the down lead construction is a more
serious problem.

The signal leaving the mixer in the outdoor 'front end'
unit, at a frequency of from 950-1750 MHz, will not
suffer very much loss in traversing the 10 metres or so
of low-loss cable that forms the familiar downlead used
already in reception of u.h.f. television. But, although
the cables used for down-leads have a fairly low
attenuation factor, they may be prone to pick up unwanted
signals that may exist at the domestic installation.
These sources of interference were listed in earlier
chapters and their impact upon the domestic scene depends
upon the proximity of airfields, etc. and may be very
localised, deserving localised treatment. However there
is already an EEC guideline, based upon an original
British Standard specification, on the anticipated leak-
age into u.h.f. cables which may be used for the purpose
of down-leads for satellite broadcast i.f. signals. In
order to 'swamp' such pick-up on the down-lead, the i.f.
level needs to be of the order of -78 dBW (2 mV e.m.f. in
a 70 ohm cable) and requiring an overall gain of around
30 dB in the front end.

5.3.4 The Indoor Unit

This is a major piece of electronics which will
undoubtedly make great demands on new technology. It
will contain the following items:

1. A filter to reject interference and noise
2. A mixer and the 2nd local oscillator
3. The second IF filter - probably a SAW
   filter (Surface Acoustic Wave filter)
4. An AGC amplifier. (An output of the AGC
   voltage may be required for antenna alignment)
5. Limiter
6. FM discriminator (demodulator). Two may be
   required, one for the vision and one for the
   sound (see later note on this subject)
7. Energy dispersal elimination circuits
8. De-emphasis circuits
9. Signal processing. In the modern receiver
   this will be largely digital
10. Coder for the vision signal depending on the
    standards employed by DBS, for example, PAL.
11. UHF modulator for vision and sound
12. Stereo sound output to drive a domestic
    Hi-Fi system
13. "Cenelec",standardised peritelevision connector
    or a similar system to allow the assembly of unit
    television to meet the specific customer
    requirements
14. A user friendly operating system to make it
    possible for those less skilled to operate
    the receiver. This may include an alpha-
    numeric display which may also be used to
    display billing information
15. De-scrambling circuits for vision and sound
    signals

230

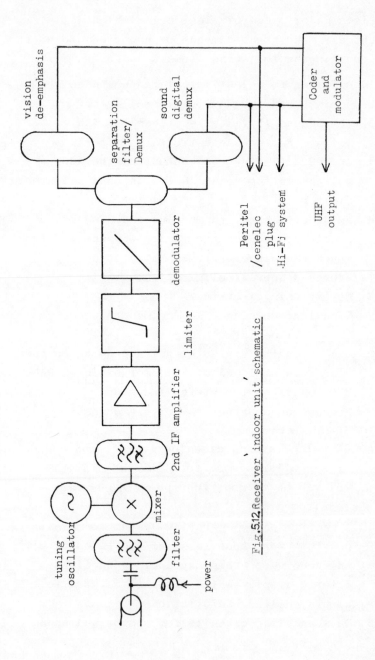

vision
de-emphasis

separation
filter/
Demux

sound
digital
demux

demodulator

limiter

2nd IF amplifier

mixer

filter

tuning
oscillator

power

Coder
and
modulator

Peritel
/cenelec
plug
Hi-Fi system

UHF
output

Fig.5.12 Receiver 'indoor' unit schematic

16. Conditional access facilities associated with
    the payment and business operation of subscrip-
    tion or "pay-per-view" systems
17. Data output circuits
18. VCR output to enable the viewer (or listener)
    to record one signal while watching another.

Clearly there will be a temptation to manufacture a low
cost version eliminating some of these facilities.
However, experience has shown that such a receiver is
then vulnerable to competition and, in that the facilities are provided by the integrated circuits, they will
be built into even the earliest sets.

Tuning Range

During the planning of the frequency assignments of the
800 MHz band from 11.7 to 12.5 GHz, the idea of making
the receiver able to tune over the whole range was
considered to place too much of a difficulty in the way
of the receiver designer in cramping the choice of
intermediate frequency band. So each country's allotted
set of channels was arranged to lie in either the upper
or the lower half of the band; then, by choosing one or
other of two specific local-oscillator frecuencies, all
the 'local' or 'national' channels could be translated
into one common i.f. band of only 400 MHz width - the
band of from 900 to 1300 MHz was suggested as being
conveniently low and high enough to clear the uppermost
u.h.f. television channel at 864 MHz.

However, since that stage of the planning, two things
have happened. First, in the U.K. the lower end of the
i.f. band has been pushed upwards to 950 MHz to avoid
break-through from the newly-agreed Citizen's Band at

930 MHz. Second, the problems of tuning over the whole 800 MHz range have not been as great as was expected, resulting in the consideration of an i.f. band of from 950 - 1750 MHz. More recent proposals suggest that, in order to avoid some of the radar interference, a band from 1410 MHz to 2210 MHz might be more appropriate (or 1410 to 1810 MHz for the half band).

The facility of covering the whole band in one receiver will allow greater flexibility by doubling the number of channels to choose from; at the same time, it may put more pressure on the designer to provide multiple outputs simultaneously and it may put pressure on the aerial designer to provide easier alteration to the polarisation of the aerial and to the redirection of its beam to another satellite.

Frequency Modulation Demodulator

It should be noted that there are two key elements in obtaining a receiver with the highest possible performance.

The first is the NF of the front end emplifier and the second is the performance of the FM demodulator.

There seem to be three alternatives for the demodulator, viz:
1. A linear frequency to voltage converter such as the well known Foster Seely discriminator
2. A delay line demodulator (perhaps implemented using a SAW filter).
3. A phase locked loop demodulator. This can create a moving window in the spectrum which excludes much of the noise but has its position adjusted to accept the required carrier signal. Providing the signal has only

	2 Demodulators (Optimum)	1 Demodulator Only		
		Linear Frequency /Amplitude Converter	Delay Line Demodulator	Phase Locked Loop
Sound	0 dB	1.5 dB	0 dB	2 dB
Vision	0 dB	2 dB	3 dB	0 dB

Fig.5.13 Relative performance of different F.M demodulators
(positive numbers represent degradation)
Based on the work of Shelswell, Wilson and Zubrzycki

limited-amplitude high-frequency components this can
give an improved signal-to-noise ratio (threshold
extension).

The designer of the receiver has the option of
providing two detectors and getting optimum performance
on both sound and vision or making a compromise and
providing one demodulator to serve both purposes. It
is to be hoped that, if a compromise is made, it will
not be to the disadvantage of the sound system which is
often the weakest element (in the C-MAC system for
example).

Typical performance of the various decoders is
illustrated by the table in Fig.5.13.

5.3.5. Scrambling and Conditional Access

The process of providing and operating a fee paying
system of broadcasting has important implications on
the programme, financial, technical and legal aspects
of the service.

It is convenient to consider the design of the system
in two parts.

First there is "scrambling". This is the technical
process which renders the signal unintelligible and so
prevents its unauthorised use.

The second is the "conditional access" system which is
the means whereby the audience is given the key to
unscramble the picture. This key may be freely available or available on subscription only (or pay-per-view).

235

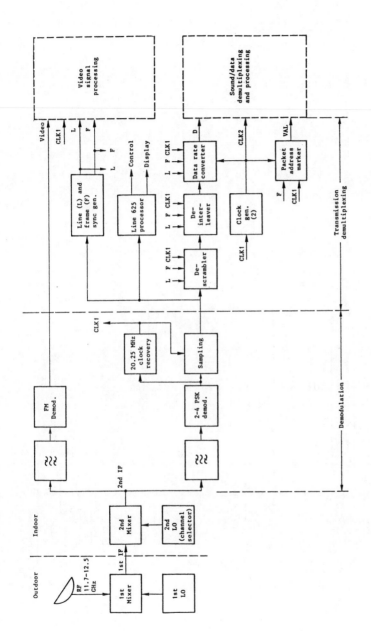

Fig 5.14 Schematic of the C-MAC Receiver (Courtesy EBU, Brussels)

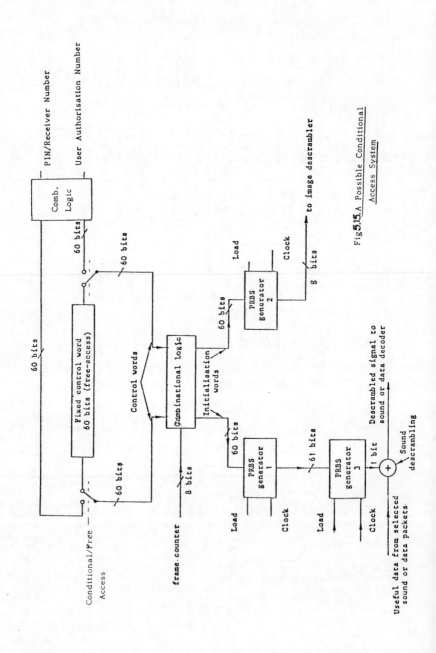

Fig 5.15. A Possible Conditional
Access System

This second part is very dependent on local factors and the business structure. It is not considered appropriate to consider here the various ways in which the money may be collected and the viewer authorised to received the programmes.

In contrast the scrambling system is of universal interest and is a very important factor in the design of the receiver. It is relatively inexpensive to implement in the digital receiver and there are those who believe that all signals should be scrambled because the resulting signal has an advantage in terms of interference by about 2 dB. This is of considerable value to countries whose objectives were not fully met by the WARC plan. In these circumstances the distinction between free access and conditional access systems is only whether the key is freely available or has to be paid for.

The scrambling process required in each country depends on the law. If protection from the law is weak then scrambling has to be strong, that is, resistant to piracy. It is considered that, in general, support from the law is poor.

Scrambling digital sound is an easy task. One has only to add a random digital word to the signal to render it unintelligible. ("modulo two" addition is perhaps the simplest process).

The most secure method of scrambling the video is "active line rotation". In this system the television line is cut in two and the two parts transposed in time. This is a relatively simple process in a receiver using digital processing and there is no difficulty in building a shift register or digital store to carry out this

operation. The position of the cut is determined by a
random number and varies from line to line.

Both of these processes have the advantage of being
transparent to the broadcast signal, that is, the
decoded signal is free from impairment arising from the
scrambling process.

Unfortunately active line rotation is incompatible with
cable systems that employ vestigial side band amplitude
modulation. This technique is liable to introduce low
frequency distortion which, although acceptable in
television pictures, becomes converted into low
frequency noise by the "cut and rotate" process. This
low frequency noise can cause a serious degradation in
picture quality. In some circumstances it may be
desirable to employ a scrambling system which is more
compatible with cable, even if this means it must be
less secure. The other alternatives are either to
unscramble the signal at the cable head or to distribute
as a private and secure FM signal.

The key to undoing both sound and vision scrambling is
a random number. This number may be the combination of
several different signals (or keys) which may be broadcast, sent by post or purchased in a shop. While the
process cannot be described in detail for obvious
reasons, it can be observed that the digital number is
usually generated by Pseudo Random Binary Sequence
generator (PRBS).

The principal part of a PRBS generator is a large
shift register which has various taps along its length.
These taps feed a signal back to the beginning and
cause a complex variation in pattern of 1's and 0's
within the binary bit stream leaving the shift register.

Thus although the digital words are not truly random
the period of the pattern is so large that they are
effectively random. The essential confidential element
needed to decode the signal is the "initialisation"
word which sets all the segments of the shift register
before the process starts.   In practice the PRBS may be
given a new initialisation word every few seconds.   This
process can be elaborated, using more than one PRBS, to
the point where, to all intents and purposes, the signal
code cannot be broken.

The initialisation word is a combination of a frame
counter and control words for the sound and vision
signals.   It could be derived from Personal Identification Numbers (or a number buried within the receiver)
and user authorization distributed by post, by telephone
or broadcast over the air.

There are of course very many simpler schemes which
disturb the signal less severely.   Individual circumstances will dictate whether it is necessary to use a
very secure system of the type described above.

The design of secure coding systems is currently being
studied in many laboratories throughout the world.

REFERENCES

5.0.0   International Telecommunications Union 1977
World Broadcasting. Satellite Administration
Radio Conference.
Final Acts. Geneva 1977
also
Radio Regulations 1979, Appendix 30

5.2.4   CCIR SG 10/11 "Green Book" pt 2
C/N and picture quality

5.3.1   Lining up a dish by solar transit
Philip Haines
Satellite Television News, 1984, p.581

5.3.4   Cenelec connector
EBU Review (Technical)
No.198, April 1983

5.3.5   Condition Access and Scrambling
EBU Review, October 1984, p. 237

Ariane Launch (photo courtesy European
Space Agency)

CHAPTER 6

The Law Relating to Satellite Broadcasting

Three Appendices are central to the discussion contained
in this chapter. They are as follows:

Appendix 12
Convention Relating to the Distribution of Programme-
Carrying Signals Transmitted by Satellite.
Brussels, 21 May 1974.

Appendix 13
Treaty on Principles Governing the Activities of States
in the Exploration and Use of Outer Space, including the
Moon and other Celestial Bodies.
London, Moscow and Washington, 27 January 1967.

Appendix 14
United Nations General Assembly Resolution
on Principles Governing the Use by States of Artificial
Earth Satellites for Direct Television Broadcasting.

6.1 DIRECT BROADCASTING SATELLITES - COVERAGE & OVERSPILL

It is in the nature of radio waves that they do not
respect national boundaries. We are accustomed to radio
and television programmes being picked up far outside the
frontiers of the state of the originating broadcaster.
This has been particularly so in the case of short wave

radio broadcasts. The advent of satellite broadcasting has however generated great interest in the effects of broadcasting by one country on the cultural, social and political life of other countries. The United Nations, UNESCO, the Council of Europe and the EEC have all expressed concern about the effects of direct broadcasting by satellite.

The 1977 WARC General Plan for Regions 1 and 3 with its emphasis on national coverage has perhaps limited, although by no means eradicated, this concern. In any event, under the ITU regulations, even for terrestrial broadcasts power must be so limited (except for short wave broadcasts) as to serve the national territory. For satellite broadcasts Radio Recommendation No. 428A provides:
"In devising the characteristics of a space station in the broadcasting satellite service, all technical means available shall be used to reduce to the maximum extent practicable, the radiation over the territory of other countries unless an agreement has been previously reached with such countries".

The desire to limit the signal to national territory unless consent has been obtained runs like a thread through the debate on direct broadcasting by satellite. Against is set the argument for the free flow of information without the interference of governments.

The debate has been going on in the United Nations since 1967 and was resolved in 1982 in favour of the proponents of "prior consent" where broadcasts are deliberately aimed at the overseas audience.

When applying this decision legal questions have to be
answered both in the public and private law sectors.
To whom do the satellites belong, what regulations apply
to their launch and use on the one side and the
conditions under which programme material is acquired
and exploited on the other.

6.2 REGULATION AND CONTROL

Space law has been developed principally under the
auspices of the United Nations Committee on the Peaceful
uses of Outer Space. In 1967 a Treaty, shortly known as
the Outer Space Treaty, was established in order to
preserve outer space as an area of peaceful use. This
"Treaty on Principles Governing the Activities of States
in the Exploration and Use of Outer Space, including the
Moon and other Celestial Bodies" provides:

Article I "The exploration and use of outer space,
including the moon and other celestial bodies, shall be
carried out for the benefit and in the interests of all
countries, irrespective of their degree of economic or
scientific development, and shall be the province of all
mankind. Outer space including the moon and other
celestial bodies shall be free for exploration and use
by all states without discrimination of any kind, on a
basis of equality and in accordance with international
law, and there shall be free access to all areas of
celestial bodies...."
Article II "Outer space including the moon and other
celestial bodies, is not subject to national appropriation by claim of sovereignty by means of use or
occupation, or by any other means".

Notwithstanding this clear statement certain equatorial
states, on the analogy of a states' claim to sovereignty

over its air space, in 1976 attempted to claim
sovereignty over the space segments above their national
territory. This would have had the effect of giving them
the right to control the entry into and use of satellites
in their space segments. This claim would seem to be in
contradiction of the terms of the Outer Space Treaty, and
subsequently certain of the states have signed the 1977
Geneva Plan.

Article VII. The Treaty goes on to provide that each
state which launches an object into space, or from whose
territory such an object is launched is responsible for
any damage which it causes to any other state or its
citizens whether the damage occurs on earth or in space.

States retain international responsibility for national
activities in space - whether carried out by governmental
or non-governmental agencies (Article VI). It is also
provided that states retain jurisdiction and control over
objects on their registry launched into space and over
any personnel thereon while in outer space or on a
celestial body and ownership of such objects or their
component parts does not change by virtue of their being
in space (Article VIII).

The question of applicable law will arise in another
form in considering copyright and performers rights in
radio and television programmes carried by DBS, but as
far as public law is concerned the Treaty makes clear
that the jurisdiction over the satellite itself is that
of the state on whose register the satellite is placed.

International Telecommunications Union rules apply to
satellite broadcasting and the Radio Regulations as
revised by WARC-ST 1971 (World Administrative Radio
Conference for Space Telecommunications) state that "in

the broadcasting satellite service the term 'direct reception' shall encompass both individual reception and community reception". The detailed definitions are as follows:

Broadcasting-Satellite Service: A radiocommunication service in which signals transmitted or re-transmitted by space stations are intended for direct reception by the general public.

Individual reception: The receptions of emissions from a space satellite in the broadcasting-satellite service by simple domestic installations and in particular those possessing small antennae.

Community receptions: The reception of emissions from a space station in the broadcasting-satellite service by receiving equipment, which in some cases may be complex and have antennae larger than those used for individual reception, and intended for use:
- by a group of the general public in one location
- through a distribution system covering a limited area

The overall legal regime for cable distributors of foreign programmes is now, after a decade of intense debate and legal activity, reasonably well settled in Europe and will be considered in relation to copyright. It would not seem of prime legal importance whether the signal is delivered to the cable head end by a satellite of whatever kind or by terrestrial transmitters.

What is clear, however, is that in considering plans for direct broadcasting by satellite in Geneva in 1977 the World Administrative Radio Conference followed the long established principle of adopting a plan based on natural reception. Any overspill is, therefore, to be regarded as technically unavoidable. In looking at the plans for DBS in Region 1 and Region 3 we are principally concerned, unless governments agree otherwise, with programming which is domestically based. This simplifies the legal problems both in the public and private law fields.

Direct broadcasting by satellite will not only be covered by the ITU regulations but also by the general considerations of space law where they are applicable. The United Nations General Assembly in 1967 asked the Committee on the Peaceful Uses of Outer Space (UNCOPOUS) to examine the whole question of DBS. The Committee's Legal Sub-Committee drafted a set of principles governing the use of artificial earth satellites for DBS with a view to formulating an International Convention. While it was possible to reach a consensus on many of the principles this did not prove possible on the most politically sensitive. For the first time in its dealings with space law the UN General Assembly in December 1982 adopted a resolution which did not have total approval. At issue was the conflict between those states which placed greater emphasis on the free flow of information and those which sought, succesfully in the result, to insert the principle of 'prior consent' to international DBS signals.

Article 19 of the Universal Declaration of Human Rights provides that:
"Everyone has the right to freedom of opinion and expression; this right includes the freedom to hold opinions without interference and to seek, receive and import information and ideas through any medialand regardless of frontiers".

This basic principle was further elaborated in Article 10 of the Council of Europe's 1954 Convention on Human Rights:

1. "Everyone has the right to freedom of expression. This right shall include freedom to hold opinions and to receive and impart information and ideas without interference by public authority and regardless of frontiers. This article shall not prevent States from requiring the licensing of broadcasting, television or cinema enterprises".

2. The exercise of these freedoms, since it carries with it duties and responsibilities, may be subject to such formalities, conditions, restrictions or penalties as are prescribed by law and are necessary in a democratic society, in the interests of national security, territorial integrity or public safety, for the prevention of disorder or crime, for the protection of health or morale, for the protection of the reputation or rights of others, for preventing the disclosure of information received in confidence, or for maintaining the authority and impartiality of the judiciary.

It will be seen that there is a basic freedom to receive and impart information and ideas without inteference by public authority and regardless of frontiers. It is questionable whether within the exceptions in paragraph 2 can be found room for an overall right of prohibition based on general protection of cultural or political values.

In any event in Region 1 and Region 3 the position is clear. Each country has been allocated up to five channels for essentially national coverage unless either a group of countries agree to join together and pool

certain of their channels to provide supra national
programming or a neighbouring country accepts programmes
specifically directed to it. In the main though we have
to consider the case of technically unavoidable overspill
rather than the supra national arrangements or programming which does not have a national basis.

The United Nations, whilst concerned with all overspill,
has been particularly concerned about programmes
conceived on an international basis. The principles
adopted by the United Nations General Assembly on 10th
December 1982, by 108 votes to 13 with 13 abstentions,
are to govern the use by nations of artificial earth
satellites for international direct television broadcasting. A Declaration by the United Nations General
Assembly does not have legal force in itself, but an
effort will be made to have the proposals incorporated
into a convention. The countries voting against the
Declaration, with the exception of the US and Japan,were
all from Western Europe.

There had been earlier debate in the Committee as to
whether the principles should apply to unavoidable
overspill or only to signals directed primarily at
another cojntry. The result would seem to reduce the
effectiveness of the principles so far as the European
plans for direct broadcast television in the 1980s are
concerned.

It is interesting in this context that reference is
only made to international direct television broadcasting.
The radio broadcasts which would be carried alongside
the television signal would not seem to be subject to the
same prior consent regime.

The two fundamental principles are set out as follows
under the heading "Purposes and Objectives" and
"Consultations and Agreements between States".

Purposes and Objectives

A. Purposes and Objectives

1. Activities in the field of international direct
   television broadcasting by satellite should be carried
   out in a manner compatible with the sovereign rights
   of states, including the principle of non-intervention
   as well as with the right of everyone to seek,
   receive and impart information and ideas as enshrined
   in the relevant United Nations instruments.

2. Such activities should promote the free dissemination
   and mutual exchange of information and knowledge in
   cultural and scientific fields, assist in
   educational, social and economic development,
   particularly in the developing countries, enhance
   the qualities of life of all peoples and provide
   recreation with due respect to the political and
   cultural integrity of states.

3. These activities should accordingly be carried out
   in a manner compatible with the development of mutual
   understanding the strengthening of friendly relations
   and cooperation among all states and peoples in the
   interest of maintaining international peace and
   security.

G. Duty and Right to Consult

10.Any broadcasting or receiving state within an
   international direct television broadcasting
   satellite service established between them requested
   to do so by any other broadcasting or receiving state
   within the same service should promptly enter into
   consultations with the requesting state regarding its

activities in the field of international direct
television broadcasting by satellite, without
prejudice to other consultations which these states
may undertake with any other state on that subject.

J. Consultations And Agreements Between States

13. A State which intends to establish or authorize
the establishment of an international direct
television broadcasting satellite service shall
without delay notify the proposed receiving state
or states of such intention and shall promtly enter
into consultation with any of those states which so
requests.

14. An international direct television broadcasting
satellite service shall only be established after
the conditions set forth in Paragraph 13 above have
been met and on the basis of agreements and/or
arrangements in conformity with the relevant
instruments of the International Telecommunication
Union and in accordance with these principles.

15. With respect to the unavoidable overspill of the
radiation of the satellite signal, the relevant
instruments of the International Telecommunication
Union shall be exclusively applicable.

It is clear, therefore, that the ITU principle of
consultation is reinforced by this text. Without the
agreement of the recipient state non-national DBS
signals may not be directed at it. States which do not
wish their cultural, social or political integrity to
be undermined by a neighbouring dominant, subversive
or ideologically opposed state would have the right to
prevent broadcasts of which they do not approve under
these principles. This could prove a serious weapon
against the policy of open broadcasting throughout the
world.

The principles further provide that international law, including the UN Charter, the Outer Space Treaty and ITU regulations, should apply.

B. Applicability of International Law

4. Activities in the field of international direct television broadcasting by satellite should be conducted in accordance with international law, including the Charter of the United Nations, the Treaty on Principles Governing the Activities of States in the Exploration and Use of Outer Space including the Moon and Other Celestial Bodies, of 27 January 1967, the relevant provisions of the International Telecommunication Convention and its Radio Regulations and of international instruments relating to friendly relations and cooperation among states and to human rights.

The principles reflect the spirit of the Outer Space Treaty with regard to equal opportunity of access to the technology, international cooperation and the peaceful settlement of disputes and of the UN role in such activity:

C. Rights and Benefits

5. Every State has an equal right to conduct activities in the field of international direct television broadcasting by satellite and to authorize such activities by persons and entities under its jurisdiction. All states and peoples are entitled to and should enjoy the benefits from such activities. Access to the technology in this field should be available to all states without discrimination on terms mutually agreed by all concerned.

D. International Cooperation

6. Activities in the field of international direct
   television broadcasting by satellite should be based
   upon and encourage international cooperation. Such
   cooperation should be the subject of appropriate
   arrangements. Special consideration should be given
   to the needs of the developing countries in the use
   of international direct television broadcasting by
   satellite for the purpose of accelerating their
   national development.

E. Peaceful Settlement of Disputes

7. Any international dispute that may arise from
   activities covered by these principles should be
   settles through established procedures for the
   peaceful settlement of disputes agreed upon by the
   parties to the dispute in accordance with the
   provisions of the Charter of the United Nations.

I. Notification to the United Nations

12. In order to promote international cooperation in
    the peaceful exploration and use of outer space,
    states conducting or authorizing activities in the
    field of international direct television broadcasting
    by satellite should inform the Secretary General of
    the United Nations, to the greatest extent possible,
    of the nature of such activities. On receiving this
    information, the Secretary General should disseminate
    it immediately and effectively to the relevant
    specialized agencies as well as to the public and the
    international scientific community.

International responsibility is placed on states for the
conduct of those carrying our international DBS
activities in space. In so far as certain states stand

back from regulating programme content such a principle
may not be universally acceptable.

F. State Responsibility

8. States should bear international responsibility for
   activities in the field of international direct
   television broadcasting by satellite carried out by
   them or under their jurisdiction and for the
   conformity of any such activities with the principles
   set forth in this document.
9. When international direct television broadcasting by
   satellite is carried out by an international intergovernmental organization, the responsibility
   referred to in Paragraph 8 above should be borne
   by that organization and by the states participating
   in it.

Finally the principles refer to the provisions which
have to be made for securing copyright and neighbouring
rights. As will be discussed under the heading of
Rights it is widely accepted that the rules which apply
to terrestrial broadcasting in these fields will
basically apply to direct broadcasting by satellite.
The more complicated legal issues in this sphere arise
in the area of point to point satellites.

H. Copyright and Neighbouring Rights

11.Without prejudice to the relevant provisions of
   international law, states should cooperate on a
   bilateral and multilateral basis for protection of
   copyright and neighbouring rights by means of
   appropriate agreements between the interested states
   or the competent legal entities acting under their
   jurisdiction. In such cooperation they should give
   special consideration to the interests of developing
   countries in the use of direct television.

6.3 COPYRIGHT

The use of point to point or distribution satellites to
convey material raised problems of protection which were
not covered by existing intellectual property conventions.
Just as the unauthorised use of literary, artistic and
musical works led to the formulation of the Berne and
Universal Copyright Conventions so the unauthorised use
of signals transmitted by satellite led to the Brussels
Satellite Convention of 1974.

Under copyright law an author has the right to control
the broadcast use of his work. Article 11(bis) of the
Berne Copyright Convention provides:

> Article 11(bis), paragraph (1)
> Scope of the Right

(1) Authors of literary and artistic works
shall enjoy the exclusive right of
authorizing:

(i) the broadcasting of their works or
the communication thereof to the public
by any other means of wireless
diffusion of signs, sounds or images;

(ii) any communication to the public by
wire or by rebroadcasting of the
broadcast of the work, when this
communication is made by an organisation
other than the original one;

(iii) the public communication by loudspeaker
or any other analogous instrument
transmitting, by signs, sounds or
images, the broadcast of the work.

The meaning of broadcasting as a communication to the
public is in line with the ITU Radio Regulations.
Broadcasting involves the despatch of signals by Herzian
waves and includes all methods of doing so. The
essential point is that no intermediary body interposes
between the emitting antennae and the aerial at the
receiving point. What matters is that the whole
operation is carried out by a single organisation.
Where the signal is rebroadcast or communicated to the
public by cable the separate author's right under
paragraph (1)(ii) is involved. The use of direct
broadcast satellites in which the signal is intended
for reception by the general public in copyright law,
therefore, is within the scope of the rules governing
broadcasting.

The Universal Copyright Convention under Article IVbis(1)
also requires that national legislation should protect
the author's exclusive right to authorise reproduction
of his work by any means, its public performance and
broadcasting.

With the development of technology it was seen that
the phonogram and the broadcast which fixed an artist's
performance and enabled it to be used without the
presence or consent of the artist constituted a threat
to the integrity and livelihood of performing artists.
Equally the skill and cost involved in the creation and
production of phonograms and broadcasts could be
nullified by their unauthorised reproduction. The Rome
Convention 1961 which conferred a "neighbouring right"
on Performers, Producers or Phonograms and Broadcasters
was an attempt to extend protection analogous to copyright protection in the technological era. In the
Convention, broadcasting is defined at Article 3(f):

'"broadcasting" means the transmissions by wireless means for public reception of sounds or of images and sounds'.

With the advent of communication satellites a still unresolved debate began as to the nature of the transmission of signals which were not intended for direct reception by the public. Protection of copyright and neighbouring rights had developed prior to these new methods of communication. In addition much of the material carried by satellite was not of a nature to be covered by classic copyright concepts - sports events, news relays for example. At the same time the very popularity and importance of these events made them targets for potential pirates.

The legal solution chosen was the formulation in 1974 of the Brussels Satellite Convention. Whereas the other conventions were in the domain of private international law, giving rights to individuals or organisations, this new Convention was to be enforced under international public law. It was left to states to decide the best means of controlling piracy in their territory. An individual right holder could have no direct redress. The Convention protects "programme-carrying signals" (and not the programme content itself) against the reception of signals from point to point or distribution satellites by organisations for which they are not intended and their subsequent unauthorised distribution to the public. Article 3 of the Convention states specifically that it "shall not apply where the signals emitted by or on behalf of the originating organisation are intended for direct reception from the satellite by the general public".

The Convention recognises the value of the signal to the broadcasters irrespective of the nature of the material which it carries. The definitions on which the Convention is based are set out in Article I:

Article I

For the purposes of this Convention:

(i)   'signal' is an electronically-generated carrier capable of transmitting programmes;

(ii)   'programme' is a body of live or recorded material consisting of images, sounds or both, embodied in signals emitted for the purpose of ultimate distribution;

(iii) 'satellite' is any device in extraterrestrial space capable of transmitting signals;

(iv) 'emitted signal' or 'signal emitted' is any programme-carrying signal that goes to or passes through a satellite;

(v)   'derived signal' is a signal obtained by modifying the technical characteristics of the emitted signals, whether or not there have been one or more intervening fixations;

(vi) 'originating organisation' is the person or legal entity that decides what programme the emitted signals will carry;

(vii) 'distributor' is the person or legal entity that decides that the transmission of the derived signals to the general public or any section thereof should take place;

(viii) 'distribution' is the operation by which a distributor transmits derived signals to the general public or any section thereof.

By Article 2 each contracting state undertakes to take adequate measures to prevent the distribution on or from its territory of any programme-carrying signal by any distributor for whom the signal emitted to or passing through the satellite is not intended. This obligation applies where the originating organisation is a national of another Contracting State and where the signal distributed is a derived signal.

With this type of programme signal it will be for the broadcasters to show the countries or organisations for which the signal was intended. What is clear is that piracy can be prevented on the 'up' or 'down' leg of the signal following words in Article 2 "emitted to or passing through the satellite".

It is up to each Contracting State to decide whether to grant protection within the framework of copyright or neighbouring rights, penal sanctions or administrative or telecommunications' law. It is also for each state to decide on the length of time for which protection is to be granted. States may also grant exemptions in favour of short extracts used for reporting current events or quotations for information purposes and developing countries may permit the distribution of programmes for the purposes of teaching or scientific research.

The Convention is not intended to derogate from any existing national legislation which protects rights holders and it was seen by the broadcasting organisations which initially promoted it as a method of achieving protection via the signal which would benefit all programme contributors through the limiting of piracy.

It has to be said, however, that the Convention has not gained wide acceptance. Only eight countries have ratified the Convention - Austria, Federal Republic of Germany, Italy, Kenya, Mexico, Morocco, Nicaragua and Yugoslavia. The United Kingdom government has stated its desire to amend national legislation so as to make ratification possible.

However, the distinction between direct broadcasting by satellite and the use of a distribution (point-to-point or communcations) satellite has important legal consequences; with the blurring of the distinction between these two categories in some parts of the world these legal differences are under pressure. The essential features of the continuing debate are as follows.

A Direct Broadcast Satellite is generally understood in legal terms as meaning a satellite capable of modifying programme-carrying signals from space for direct reception by the public. A Distribution Satellite is understood to mean a satellite transmitting programme-carrying signals to be modified for public reception by a suitable earth station. As stated earlier broadcasting itself is understood in copyright law as meaning tele-communications of sounds and/or images by means of radio waves (electromagnetic waves of frequencies lower than 300 GHz) for reception by the public at large. For the purposes of the Rome Convention for the Protection of Performers, Producers of Phonograms and Broadcasting Organisations it means the transmission by any wireless means for public reception of sounds or of images and sounds.

The legal debate, therefore, centres on the concept of "public reception". The reception of programme-carrying signals from a satellite without the intermediary of an earth station transforming the emitted signals into radio waves capable of being received by the public would come within the ambit of broadcasting. The use of the Direct Broadcast Satellite which itself transforms the signals in in legal terms, therefore, analogous with terrestrial broadcast transmitters. Clearly questions of space law arise but the DBS can be comprehended in conventional copyright terms.

This, according to some authors, is not the case with distribution satellites.

Both the Universal Copyright Convention and the Berne Convention give an author the exclusive right to authorise the broadcasting of his work. The Berne Convention gives the more developed right but inevitably there is debate over the meaning of "broadcasting" in this Convention. This definition is clearly of importance both to authors and to the owners of neigh-bouring rights under the Rome Convention. The authors have tried to argue in a number of ways.

(1) They claim a right of injection (i.e. control over the material put into the signal on the up-leg.

(2) They claim an extended definition of broadcasting which covers all stages in the passage of the signal via both DBS and distribution satellites on the ground that satellite transmission is a mere technical extension of the broadcast

and

(3) It is claimed that the concept of 'ultimate reception'

by the general public should be imported into the definition.

The importance in practical terms of the authors' concern is in relation to the use of distribution satellites to feed cable systems. If the "broadcast" takes place at the cable head it is feared that the rights holders will not be able to obtain remuneration from the satellite operator who may be the point at which advertising revenue is collected. In addition the widespread distribution of the satellite signal could involve the rights holders in seeking to pursue cable operators in countries with a low level of copyright protection and with whom they have no direct contact.

Whilst it is possible for telecommunication law, copyright law and national media law to be at variance the extended view of these copyright experts would create confusion in the telecommunication sphere.

The following points have been used in this debate.

(1)   Under the Radio Regulations reception by the general public is inherent in the concept of broadcasting. (This extract of Section 6.1 is repeated for convenience).

Broadcasting-Satellite Service is a radio-communication service in which signals transmitted or retransmitted by space stations are intended for direct reception by the general public (3.18). Reception can be both 'individual' and 'community' but both are covered by "direct reception".

Individual Reception is the reception of emissions from a space station in the broadcasting satellite service by simple domestic installations and in particular those possessing "small antennae".

Community Reception is the reception of emissions
from a space station in the broadcasting-satellite
service by receiving equipment which in some cases
may be complex and have antennae larger than those
used for individual reception and intended for use
by a group of the general public at one location
or through a distribution system covering a limited
area.

(2) Allocation of frequencies

In general the frequency allocations provide for
two separate services:

(a) the Broadcasting-Satellite Services
(b) the Fixed Satellite Services

It is argued by those who seek the extended interpretation of broadcasting that the allocation of
frequencies, particularly in Region 2 (North
America and Latin America) are not entirely
separated and are at certain points co-extensive.
Consequently from the technical point of view the
services may be hybrid so that different copyright
considerations could not apply.

(3) Secrecy of Signals under ITC

It is the responsbility under the International
Telecommunications Convention for satellite
operators to ensure the secrecy of the signals that
they carry on communications (point-to-point) and
distribution satellites, but clearly not for DBS
transmissions.

6.4 International organisations

6.4.1 INTERNATIONAL TELECOMMUNICATIONS UNION

The ITU is a specialised agency of the United Nations. 157 countries are members of the ITU and its agreements, when ratified by member states, have the force of international treaties.

The ITU harmonises the activities of the member nations with respect to telecommunication services, fosters cooperation and technical development.

The International Telecommunications Convention is an international convention to which all member states subscribe. The Plenipotentiary Conference meets every five years and its Administrative Conferences make the regulations on technical standards.

6.4.2 INTELSAT

Intelsat is the main organisation owning and operating communication satellites. Membership is open to all countries of the ITU on an intergovernment basis. Members agree that they will so regulate their regional and domestic satellites in a way which does not inter-fere with or damage the Intersat system.

6.4.3 OTHER BODIES

The United States is represented in international communication satellite matters by Comsat. There are also various groupings of the PPT's for example,Eutelsat, Arabsat, Afrosat, Nordsat.

6.5 National UK Organisation

The basis of UK broadcasting is the Wireless Telegraphy
Acts 1949 to 1967 and the Post Office Act 1969 (These
Acts being revised by a new Telecommunications Act), the
Television Act 1981, the Charter of the BBC and the
Licenses granted to the BBC and the IBA.

The responsibility for administering this legislation is
centred on:

1. The Secretary of State for Trade and Industry who is
responsible, after consultation where appropriate with
the Home Secretary, for
   (a) International policy and planning of the Radio
Frequency spectrum for broadcasting
   (b) National planning of the radio frequency spectrum
for Radio and Television networks and satellite broadcasting
   (c) The ITU (CCIR) and other technical fora
   (d) Radio Interference questions

2. The Secretary of State for the Home Office who is
responsible, after consultation where appropriate with
the Secretary of State for Trade and Industry, for:
   (a) Policy relating to the use by broadcasting
services of the frequency bands allocated nationally for
broadcasting
     Licensing of all broadcasting activities for
general reception
     The setting (and collection) of fees and charges
for all broadcasting applications for general reception
   (b) Technical standards for broadcasting applications.

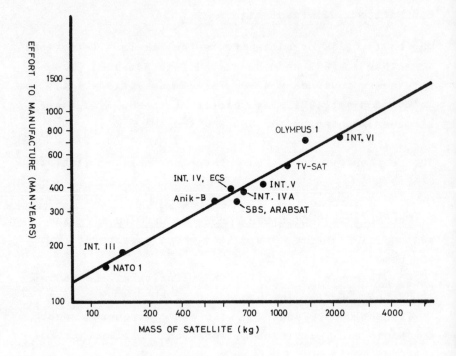

FIG.7.1 MASS AND EFFORT TO MANUFACTURE FIRST
FLIGHT UNIT OF COMMUNICATION SATELLITE
SYSTEMS

CHAPTER 7

Direct Broadcasting Satellite System Costs

7.1 ALTERNATIVE WAYS OF ESTIMATING

The price of a satellite or the services that it offers
is a bargain struck between the buyer and the seller.
The buyer needs to specify his requirements most carefully and study competitive bids to supply these
requirements.

Costs will depend on circumstances and a detailed
analysis of cost would not be appropriate in this
general discussion. In any case disclosure of precise
cost would be regarded by some people as a breach of
confidence.

Nevertheless the cost of a satellite cannot be ignored.
It is an essential element in the debate whether or not
to provide a service by satellite. It also has a vital
role in deciding whether or not to opt for a high power
or low power service. With this in mind the authors
have sought trends. If some basis for estimating can be
found then this will suffice to give the order of cost
and permit these fundamental decisions to be made.

7.1.1. Satellite costs as a function of mass (& effort)

In earlier chapters the 'custom built' nature of
satellite design has been discussed. We have seen that
different applications with different coverage areas

267

will require different transponder powers. The cost of
any communication satellite increases as its size goes
up and there is a close relationship between the cost
of a satellite and its mass. A larger satellite
requires more effort to design and construct and Fig 7.1
illustrates the relationship between the effort (in man-
years) and the mass of a satellite. It is clear that
the effort is closely related to the mass and, as this
effort is the major element in the production costs,
we can deduce that satellite mass is a good measure of
the cost of a satellite. The examples quoted in the
figure exclude the weight of fuel but include both direct
broadcasting and low power communication satellites.

> The satellite cost per kg of payload is of the order
> £150,000 (1983).

7.1.2. Launch costs

The launch cost of a satellite is also determined by its
mass, although this time we must include the total payload including the fuel necessary for station keeping.
(It should be remembered that, if we ignore equipment
failure, the amount of fuel carried by the satellite
determines its life and hence the return on the capital
invested).

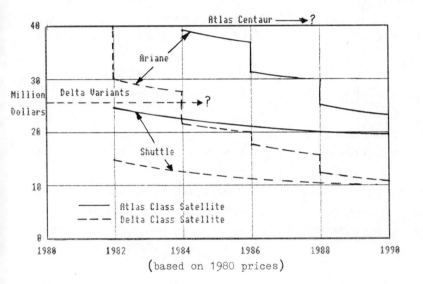

Fig 7.2   Forecast Launch Cost

The launch costs appear to be more difficult to estimate.
In Fig.7.2 there is a marked 'learning' curve. The real
cost of a launch is likely to fall as the years go by.
Furthermore it will be seen that the launch cost is very
dependent on the vehicle employed. The Atlas and Delta
rocket launches have been largely superceded by the
Shuttle and the Ariane series of rockets. In the face
of competition the Shuttle costs are likely to reduce
when the initial investment has been recovered. Equally
the Ariane costs are likely to reduce in steps as each
new and larger version of the rocket is produced.

7.1.3. System costs as a function of power

The mass of a satellite includes mass of the solar array
and is influenced by the size and number of transponders
carried. It might be suggested that the cost of a
satellite system should be related to the power required.
Fig 7.3 takes a number of examples and compares the
system cost with total power.

As an example we consider a travelling wave tube giving
about 230 watts of 12 GHz power. (The usable output
from such a tube would depend on the internal losses
within the antenna system etc, but is likely to be about
180 watts). This tube might be, say, 42% efficient.
The efficiency of the power supply providing this power
at the right potential is likely to be 90% and so, if

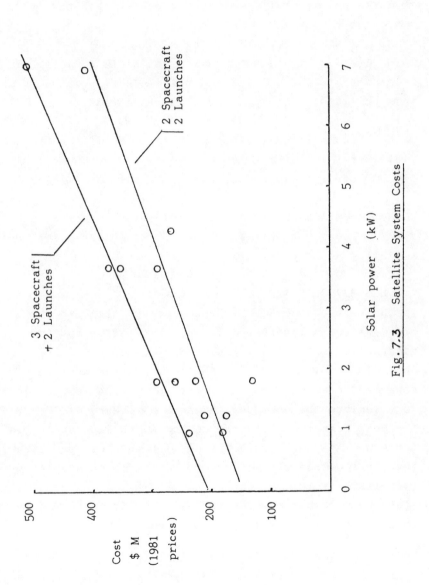

Fig. **7.3**   Satellite System Costs

the satellite carries two transponders, the power
required is about 1058 watts.

In addition the receivers on board may require, say,
20 watts and the platform "bus" carrying this payload
could demand 200 watts.

This gives a total power requirement of 1278 watts. It
is prudent to add 10% for minor items not included in
this estimate. The total power is then 1406 watts.

On this basis the satellite cost would be of the order
£18,500 per watt. The system cost would be upwards of
£80,000 per watt depending on the system specification.

The correlation of system cost and power is not good but
Fig 7.3 shows that a reduction in transponder power does
not bring about a dramatic reduction in system costs.

7.2 ECONOMY OF SCALE

In the case of the direct broadcasting satellite the
broadcaster is interested in the competitive cost of
alternative satellites. He needs to know how many
channels it will provide. What will be the available
power and what will be its lifetime? It is instructive
to calculate the cost per channel, per year. In Fig 7.4
the development cost (expressed, as discussed earlier, in
terms of man-years) is compared with the cost per circuit
year on the Intelsat series of communication satellites
(Intelsat 2 to 6). Although these are low power satellites the conclusion which may be drawn from the figure is
equally applicable to Direct Broadcasting Satellites.

During the period of development of the Intelsat system
there has been continuous improvement in technology and,
as with the cost of launching the satellite, this

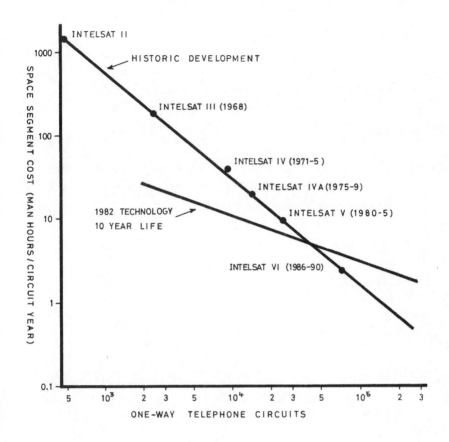

FIG.7.4 COST PER CIRCUIT FOR COMMUNICATION
SATELLITES

distorts the statistics. Fig.7.4 includes a line which gives a corrected view of the economies of scale.

To further illustrate the effect of economy of scale in DBS systems in a less theoretical manner four illustrations of practical broadcast coverages have been chosen. They are:-

1. A four channel satellite covering the smallest service area planned at the World Administrative Radio Conference in 1977

2. A medium sized satellite able to provide two broadcast channels to a small country about the size of the U.K.

3. A larger satellite providing six channels to a small country

4. The same larger satellite supplying three DBS channels to a medium size country.

These four cases are detailed in tabular form in Table7.1 and the cost per channel, per year, per square kilometre is calculated. In the table the somewhat arbitrary, but nevertheless realistic, costs given illustrate the increase in efficiency produced by using the largest satellite for which paying users can be found. The sharing of common resources in this way may be achieved by countries assigned the same orbital location sharing a satellite system equipped with separate transponder systems and antennas for each area.

TABLE 7.1 WARC '77 STANDARD-DIRECT BROADCASTING COSTS

Circular beam size	$0.6^{o}x0.6^{o}$	$1.2^{o}x1.2^{o}$	$1.2^{o}x1.2^{o}$	$0.6^{o}x0.6^{o}$
Area Covered (sq.km)				
- equatorial country	200,000	800,000	800,000	1,400,000
- latitude 50° country	400,000	1,600,000	1,600,000	2,000,000
Number of Channels	4	2	6	3
Total Radio Frequency Power	176	400	1200	1200
Total Spacecraft d.c. Power	730	1300	3550	3500
Spacecraft approx cost each	£ 20m	£ 26m	£ 50m	£ 50m
Launch approx cost each	£ 10m	£ 15m	£ 30m	£ 30m
Total investment cost	£ 95m	£123m	£230m	£230m
Cost/channel/annum /sq.km covered (latitude 50° country)	£ 11.60	£ 7.10	£ 4.00	£ 4.30

Notes

1. Similar EIRPs and output efficiencies are assumed for missions

2. Circular beams are assumed but elliptical, covering the same land area, are equivalent

3. Total investment assumed as 3 satellites, 2 launches, an earthstation (£10m) and launch insurance (£5-10m/launch)

4. Lease of investment assumed to require annual payments of 1/7th of capital value per annum plus annual system running costs (£5m p.a.)

5. Elevation angles at centre of coverage of 50° and 25° are assumed respectively for equatorial and high latitude countries

6. Coverage calculated by (2 x range x sin(0.5 x beamwidth)squared x cosec elevation angle

TABLE 7.2   ASSUMPTIONS FOR DIRECT BROADCASTING SATELLITE
SYSTEM COMPARISON

	WARC '77 DBS	Low Power Broadcasting
Number of channels per satellite	6	9
Transmitter power	200 watts	20 watts
Cost/channel/annum	£6.3m	£2.0m
Receiver antenna diameter	0.9 metre	2.9 metres
Receiver cost	£ 250	£ 1000

TABLE 7.3 SYSTEM COMPARISON

Number of receivers	10,000,000	1,000,000	100,000
WARC '77 SYSTEM			
Receiver cost (£m)	2,500	250	25
5-channels x 10 years (£m)	315	315	315
TOTAL (£m)	2,815	565	340
LOW POWER			
Receiver cost (£m)	10,000	1,000	100
5-channels x 10 years (£m)	100	100	100
TOTAL (£m)	10,100	1,100	200
Ratio (WARC'77:Low Power)	1 : 3.6	1 : 2.0	1.7 : 1

7.3 EFFICIENCY

Direct broadcasting satellite coverage areas designated
by WARC '77, being elliptical in shape, vary widely in
the efficiency with which they illuminate the target
service area. Geographic and political factors seldom
produce geometrically shaped countries and, since the
satellite beam is designed to cover everywhere inside
the territory, adjacent areas are inevitably illuminated.
An estimate shows that the relation between the assigned
elliptical coverage area of the UK and the actual target
area gives an illumination factor of 15%. For some
countries, particularly those made up of a number of
islands, the illumination factor is lower than the UK;
for many others, larger.

The television-household density is a crucial factor in
determining what is the optimum satellite broadcasting
solution. The higher-power signals planned by WARC '77
are best suited to the highly industrialised parts of the
world. Furthermore, these are mainly situated in the
low-rainfall temperate zones for which this plan's
12 GHz frequency band is well suited.

Sparsely populated and developing parts of the world,
where the number of television receivers per square
kilometre is low, can use lower power and lower frequency
satellite broadcasting. The optimum national arrangement
is found by considering overall costs of the satellite
system and the receivers working with it. Of course,
this idealised approach neglects the important fact that
the satellite system and domestic receivers may be paid
for in different ways. Nevertheless the overall
considerations are a good starting point and may best be
illustrated by the comparison given in Tables 7.2 & 7.3.

The comparison given in these tables is for a country
about the area of the UK (beam 1.2^O x 1.2^O) and the
highest number of receivers corresponds to about 50% of
the UK television viewers taking advantage of satellite
broadcasting. The UK's average population density is
about 100 times that of the most sparsely populated
areas of the world and therefore the two other columns
illustrate the situation either where population is at
a lower density or where the number of receivers per
head of population is below that obtained in industrialised societies (about 1 television set for every 3
people).

The WARC '77 satellite system costs are those quoted
earlier for a 6-channel satellite and the Low Power case
is based on the same costs as those used for the 0.6^O x
0.6^O beam DBS satellite but equipped with even lower
power transmitters. The comparison clearly shows that
the WARC '77 system is cost effective at high and medium
receiver densities and only becomes uneconomic at very
low densities. Such low receiver densities could
correspond to the idea of village-community reception
in developing countries. For this application, use of
the 2 GHz broadcasting band would lower the cost of
receivers and the antenna. This could partially off-set
the higher performance required of them by the low
power solution.

7.4 CONCLUSIONS

It was not thought appropriate to give specific costs
and an attempt has been made to obtain general information which is of value when considering many different
applications.

The figures obtained are confused by the passage of time.
Over a period of years there has been inflation and
improvements in technology which distort comparisons.
It is customary to remove the effects of inflation by
referring to a specific year but it is not as easy to
allow for the development of technology. As a result,
the statistics available are insufficient to give an
accurate picture.

It should also be noted that an accurate comparison of
costs can only be made when the services offered are
identical. An allowance has to be made for insurance,
the operating costs (telemetry and command), the ground
station and cost of servicing money that is borrowed.

One may however conclude that for most countries, the
WARC plans represent an economic solution and, providing
a system is designed to take maximum advantage of the
economics of scale, a DBS service can be obtained for
£4 - £5 per television channel per year per square
kilometre.

This is then cheaper than terrestrial transmitter
coverage (in say Bands IV and V) when the costs are
calculated in a similar manner. Furthermore, the sum
involved is only a small fraction of the cost of
providing programmes on these DBS channels.

REFERENCE

Economics of Satellite Communication Systems
W.L. Pritchard
30th Congress 1979
International Astronautical Federation
IAF-79-A-09
Pergamon Press

Appendices

DERIVATION OF THE ROCKET EQUATION

Thrust force is given by the rate of change of momentum imparted by the rocket fuel. If m_e is the mass of the rocket exhaust and v_e its velocity then:-

$$\text{Thrust} = \frac{d}{dt}\, m_e\, v_e = v_e\, \frac{dm_e}{dt} \quad \text{because } v_e \text{ is constant}$$

If V is the vehicle velocity at any time t, applying Newton's law of motion

$$\left(m_d + m_f - t\, \frac{dm_e}{dt}\right) \frac{dV}{dt} = v_e\, \frac{dm_e}{dt} \quad \text{where } m_f = \text{mass of fuel}$$
$$\text{and } m_d = \text{dry mass of rocket}$$

But $\dfrac{dm_e}{dt}$ is a constant, say c, then:-

$$(m_d + m_f - ct)\, \frac{dV}{dt} = cv_e$$

This may be rearranged to give:-

$$\int \frac{dV}{cv_e} = \int \frac{dt}{(m_d + m_f - ct)}$$

which integrates to:-

$$\frac{V}{cv_e} = \frac{-\log_e (m_d + m_f - ct)}{c} + k$$

where k is another constant. This simplifies to:-

$$V = -v_e \log_e (m_d + m_f - ct) + k$$

When $t = 0$ then $V = V_0$ the starting velocity (zero for first stage)

therefore $\quad V_0 = -v_e \log (m_d + m_f) + k$

and $\quad k = V_0 + v_e \log (m_d + m_f)$

If this result is substituted in the main equation then:-

$$V = V_0 + v_e \log_e (m_d + m_f) - v_e \log_e (m_d + m_f - ct)$$

which re-arranges as:-

$$V = V_0 + v_e \log_e \frac{(m_d + m_f)}{(m_d + m_f - ct)}$$

The definition of specific impulse (I) is:-

$$\frac{\text{Thrust force}}{\text{Weight/sec. of fuel used}} \qquad \text{or:-}$$

$$I = \frac{\dfrac{d}{dt} m_e v_e}{g \dfrac{d}{dt} m_e} = \frac{v_e}{g}$$

Substituting this result for v_e and putting $m_d + m_f = M$ and $m_d = m$

$$V = V_0 + I.g. \log_e \frac{M}{m}$$

CALCULATION OF ELEVATION AND AZIMUTH TO SATELLITE

The calculation of elevation and azimuth, and also range, to the satellite, requires the use of two equations for spherical triangles and the 'cosine' and 'sine' formulae for plane triangles. Spherical triangles are made up of 'straight' lines drawn on the surface of a sphere and many textbooks describe the relations between the angles at the corners of the triangle on the sphere's surface and the angles subtended at the centre of the sphere by the sides of the triangle. The equations will be quoted, but not proved.

Such a triangle is shown in Fig 1.7, which is a perspective view of the earth showing the equator, north and south poles and a meridian connecting the north pole to the location of an earth-station at E, the intersection of the meridian and the equator at M and the south pole. The spherical triangle with which we are concerned is made up of lines on the earth's surface. The sides are a section of the meridian EM, a section of the equator MS and the 'great circle' line joining E and S. Angle EMS in the triangle is a right angle.

S is a point on the straight line joining the centre of the earth O to the location of the geo-stationary satellite P. S is called the sub-satellite point and is drawn again in Fig 1.8 which shows the plane (i.e. flat) triangle OEP. The earth's surface is also drawn in this figure and so too is a dotted line denoting the apparent line of the horizontal at the earthstation. θ is the angle of elevation at the earthstation. The side of the triangle OE is the earth's radius R (also = OS) and SP is the height, h, of the satellite above the earth's surface at the sub-satellite point.

From geography we know two of the angles subtended at the centre of the earth by the sides of the spherical triangle. One of them, ϕ in the figure, is the latitude of the earthstation at E and the other, λ in the figure, is the difference in longitudes of the sub-satellite point and the earthstation. We have to use a formula, which will be found in geometry textbooks, to work out the third angle and the relation is:

$$\cos \lambda = \cos \phi \cos \lambda + \sin \phi \sin \lambda \cos EMS$$

For the particular triangle under study, angle EMS is a right angle and, as cos 90 = 0, the equation simplifies to:

$$\cos \lambda = \cos \phi \cos \lambda$$

The angle γ is included in the plane triangle OEP and, using the 'cosine' formula gives the slant range d:

$$d^2 = R^2 = (R + h) - 2R(R + h) \cos \gamma$$

writing in the result of the spherical triangle calculation gives:

$$d^2 = R^2 + (R + h) = 2R(R + h) \cos \Phi \cos \lambda$$

This formula can be simplified to give the square of the slant range as:

$$(\text{Slant range})^2 = h^2 + 2R(R + h)(1 - \cos \Phi \cos \lambda)$$

Continuing to analyse the plane triangle OEP, this time using the 'sine' formula, we can derive the angle OEP which is equal to $90 + \Theta$ and allows calculation of the elevation angle Θ to the satellite from the earthstation:

$$\frac{R + h}{\sin(90 + \Theta)} = \frac{d}{\sin \lambda}$$

therefore:

$$\sin(90 + \Theta) = \frac{R + h}{d} \sin \lambda$$

remembering that $\cos^2 \lambda + \sin^2 \lambda = 1$ and $\cos \Theta = \sin(90 + \Theta)$

then:
$$\cos \Theta = \frac{R + h}{d} \sqrt{(1 - \cos^2 \lambda)}$$

and substituting the value for λ derived from the spherical triangle gives the elevation angle formula:

$$\cos(\text{Elevation}) = \frac{R + h}{d} \sqrt{(1 - \cos^2 \Phi \cos^2 \lambda)}$$

note that the slant distance, previously calculated, is needed to evaluate the Elevation angle.

The azimuth angle to the satellite from the earthstation is measured on the earth's surface between the line to true north and the line to the sub-satellite point. Looking again at the spherical triangle we see that this angle is 180-angle MES. We can find angle MES by using a second spherical geometric formula:-

$$\frac{\sin MES}{\sin \lambda} = \frac{\sin EMS}{\sin \lambda}$$

but angle ENS is a right angle and therefore sin EMS = 1

Re-arranging and substituting, as before, the result of the spherical triangle calculation we get:

$$\sin MES = \frac{\sin \lambda}{(1-\cos^2\phi \, \cos^2\lambda)}$$

The angle of azimuth defined above is equal to 180-angle MES and therefore the azimuth angle is given by:

$$\sin (Azimuth) = - \frac{\sin \lambda}{(1-\cos^2\phi \, \cos^2\lambda \,)}$$

Basic Earth Station-Satellite Geometry

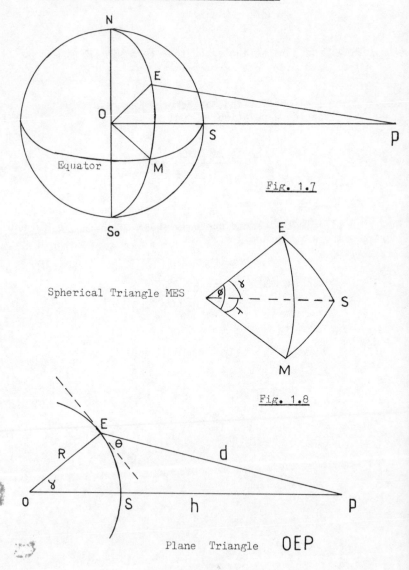

Fig. 1.7

Spherical Triangle MES

Fig. 1.8

Plane Triangle OEP

APPENDIX 3

BASIC PROGRAM TO GENERATE AZIMUTH, ELEVATION & RANGE
FOR A GEO-STATIONARY SATELLITE

```
10 input"start long.diff";s1
20 input"start lat es";s2
30 input"long diff increments";i1
40 input"lat es increments";i2
50 input"last long diff";e1
60 input"last es lat";e2
70 r=6378160
80 h=35786300
90 .for f = s2 to e2 step i2
100 for l = s1 to e1 step i1
110 l1=l*.017453293
120 f1=f*.017453293
130 d=sqr(h↑2+2*r*(r+h)*(1-cos(f1)*cos(l1)))
140 z=sqr(1-cos(f1)↑2*cos(l1)↑2)
150 x=(r+h)*z/d
160 e=int(100*57.29577951*atn(sqr((1-x↑2)/x↑2)))/100
170 y=-sin(l1)/z
180 a=int(100*(180-sgn(l1)*(57.29577951*atn(sqr(y↑2/(1-y↑2))))))/100
190 print f,l,d,e,a
200 next l: next f
```

Effective Area of Solar Cells on Spin-Stabilised Satellite

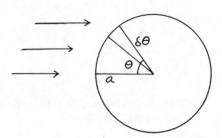

The diagram shows a section through the solar array of a spin-stabilised satellite of radius a, length l.

Consider the incident illumination to be normal to the axis of the array. Then the effective area of a strip of cells Φ from normal incidence of width $\delta\theta$ is:

$$al \cos \theta \, \delta\theta$$

The total effective area on the half of the solar array receiving any sunlight is then:

$$\int_{-\frac{\pi}{2}}^{\frac{\pi}{2}} al \cos \theta \, d\theta$$

which on integration gives:

$$\left[al \sin \theta \right]_{-\frac{\pi}{2}}^{\frac{\pi}{2}}$$

Putting in the limits gives the total effective area as 2al.

However if the solar array were all normal to the illumination then the effective area would be equal to the actual area 2 π al

Therefore the solar array of the spinning satellite is less effective than a flat array of the same area by a factor π i.e. it is less than 1/3 of the efficiency of a flat array.

The Ammonia Line   **APPENDIX 5**

The second harmonic of some currently assigned 12 GHz broadcasting
satellite transmissions will fall in the band 23.6 - 24.0 GHz;
this band is allocated solely to radio-astronomers for observing
a group of three spectral lines which are radiated by the ammonia
molecule, NH_3, in interstellar space. These studies lead to a
better knowledge of the structure and motion of our Galaxy, distant
magnetic fields and the extra-terrestrial formation of organic
molecules.

Taking the worst case, the point of interest of the radio astronomer
may be close to the equatorial plane of the earth and right behind
the geo-stationary satellite orbit. A group of satellites all at
the same nominal orbital position, may lay down a power flux-density
of up to -93 dB (Wm^{-2}) on each of ten adjacent 20 MHz television
channels from 11.8 - 12.0 GHz. All these transmissions may be
directed to a group of countries around the radio-astronomer and
fall right down the axis of his radiotelescope, which may have an
effective gain of 70 dB. His 24 GHz receiver may be sensitive to
a change in effective antenna temperature of 5.10^{-5} degrees Kelvin,
or about 0.1% after integrating its noise output for about half an
hour. Interference at a level ten times smaller than this is
considered 'harmful' and should be prevented.

From these figures, CCIR Report 224-3 (Rev.76) rightly shows that
the 'harmful' power flux-density (p.f.d.) is -146 dB (Wm^{-2}) for a
radiotelescope of 0 dB gain and therefore -216 dB (Wm^{-2}) for a
70 dB gain. This is 123 dB below the maximum p.f.d. of one
satellite broadcast transmission and 133 dB below ten of them from
11.8 to 12.0 GHz.

The required 133 dB difference in flux-density will be the sum of
several contributions from parts of the satellite output circuit.
Of these, the travelling-wave-tube amplifier, harmonic absorber
(if used) and antenna may contribute as little as 30 dB. However,
channel-separation filters will be essential in order to avoid
intermodulation when up to five output signals are multiplexed in
the same antenna. Theoretically, a simple design of multiplexing
filter would supply the remaining 103 dB attenuation; in practice,
the filter engineering would need specific attention at harmonic
frequencies.

The worst case has a low probability and although an interesting
object may be behind the geostationary orbit, an offending satellite
will move past it at 15 degrees per hour, so reducing the integrated
harmful effects. Besides, radiotelescopes' axes are often geo-
stationery and can be set to scan the passing sky in the equatorial
plane in gaps between any offending satellites. If the band 36.4 -
36.5 GHz became cleared for radioastronomers to observe the
"excited-hydrogen"-line at 36.466 GHz, a similar restriction would
concern the third-harmonic radiation from satellite transmissions
between 12.133 and 12.167 GHz.

Interference from Uplink into Terrestrial Link

Re-arranging the previously used equation for signal strength at a distance:

$$F \text{ watts / sq.m.} = \frac{P \text{ watts} \times G}{L \times S \times A}$$

where: P is the power of the transmitter
F is the achieved signal strength
L is the output loss factor
G is the antenna gain factor in the direction of
                the receiver
A is the average atmospheric loss factor
S the spreading factor depends on earthstation
    to receiver distance d (in metres) and is $4\pi d^2$
    as before

If it is assumed that the range, output and atmospheric losses and polarisations are the same for both uplink and terrestrial transmitters then the comparison between the received signals is:

$$\frac{F \text{ watts/sq.m. (Terrestrial)}}{F \text{ watts/sq.m. (Uplink)}} = \frac{P\,(T) \times G\,(T)}{P\,(U) \times G\,(U)}$$

The gain factor for a 1 metre antenna of average efficienty at 18 GHz is about 20,000. The gain factor for an earthstation 30° from its main direction of transmission is 0.3. The powers are 190 watts and 0.5 watt respectively. Therefore the ratio of wanted to unwanted signals is:

$$\frac{20,000 \times 0.5}{190 \times 0.3} = 176 \quad (\text{is } 22.5 \text{ dB})$$

The limit of acceptable interference varies with modulation details and other factors but a power ratio of 1000 (30 dB) is usually required. Therefore the uplink station causes serious interference to the terrestrial system.

Interference from Terrestrial Link into Satellite Uplink

Using the equation for signal strength at a distance:

$$F \text{ watts/sq.m.} = \frac{P \text{ watts} \times G}{L \times S \times A}$$

where:- P is the power of the transmitter
F is the achieved signal strength
L is the output loss factor
G is the antenna gain factor in the direction of
the receiver
A is the average atmospheric loss factor
S the spreading factor depends on transmitter to
receiver distance d (in metres), and is $4\pi d^2$
as before

If it is assumed that the output and atmospheric losses and
polarisations are the same for both downlink and terrestrial
transmitters then the comparison between the received signal
strengths is

$$\frac{F \text{ watts/sq.m. (Terrestrial)}}{F \text{ watts/sq.m. (Downlink)}} = \frac{P(T) \times G(T) \times S(D)}{P(D) \times G(D) \times S(T)}$$

If we take the case of the UK direct broadcast signal previously
discussed with a power of 185 watts and an edge of beam antenna
gain factor of 11555 suffering interference from a terrestrial link,
as before, using half a watt into a 2 metre diameter antenna at,
say, 40 mk range, we get (using a frequency of 12 GHz which gives
the terrestrial transmit antenna gain as 34700):-

$$\frac{F(T)}{F(L)} = \frac{0.5 \times 34700 \times 1.9 \times 10e16}{185 \times 11555 \times 2 \times 2 \times 10e10} = 3855$$

(The factor 2 after 11555 is introduced to convert from signal
strength at the edge of the beam to that at the centre).
This indicates that the unwanted signal strength is 3855 times
more powerful than the wanted one.

However, in the case under consideration, the directional
discrimination of the receiving earthstation is an additional
factor. If the wanted satellite signal is assumed to be incident
on the earthstation from 30^0 elevation and the unwanted terrestrial
energy is horizontal and directly in line with the earthstation's
azimuth direction then the gain factor from the table given in

Chapter 2 is approximately = 1. The gain factor at the centre line
of the receiving system, again from the table, is 7000. The
earthstation directional discrimination therefore reduces the
wanted to unwanted signal ratios input to the receiver from
1 : 3855 to about 2 : 1. This theoretical result is, of course,
totally unacceptable.

The calculation explains why it is often considered important to
use a frequency band for satellite reception which is not also used
for terrestrial links. However, in practice, it is often found
possible to pick up satisfactory signals from a satellite in a
shared frequency band. This is because no account has been taken
of local screening effects which can easily discriminate strongly
against terrestrial signals, which gain strength with height above
the earth's surface and satellite signals which, because of their
angle of arrival, do not. It is also true that it would be a
particularly unfortunate coincidence if the terrestrial link was
absolutely aligned with the satellite earthstation as the
calculation assumes.

APPENDIX 8

Times of conjunctions of Sun and Geostationary Satellite

1. Write
$$r = 6.4 \ 10^6 \ m \quad \text{(Earth radius)}$$
$$R = 42 \ 10^6 \ m \quad \text{(Orbit radius)}$$
A = latitude of observer, degrees; North positive
$Longobs$ = longitude of observer, degrees $\left.\right\}$ East positive
$Longsat$ = longitude of satellite, degrees
$B = Longobs - Longsat$, degrees
$C = \sin A$
$D = \cos A \cos B$
$E = \cos A \sin B$

2. Find the celestial declination of the satellite

$$\delta = \arctan\left(\frac{-rC}{\sqrt{(R-rD)^2 + (rE)^2}}\right)$$

and the angle between the observer and the Earth's axis as seen by the satellite

$$\beta = \arctan\left(\frac{rE}{R-rD}\right)$$

3. Look up ephemeris tables in an Almanack (e.g. Whitaker or Reed) to find the dates on which the Sun has the same declination as the satellite. Note that, in the Northern hemisphere, the sign of A is positive but the declination is negative.

4. From the same tables, find the Equation of Time for those two days, T_e. The Equation of Time is the difference between GMT and sundial time at the Greenwich meridian. But note carefully that Whitaker and Reed have chosen opposite sign conventions for their tables. As a safeguard against an error in sign, remember that the Sun is late in February and July, and early in May and November.

5. The mean time of the conjunction is made up of three parts :-

$$T_m = (\pm)T_e + 4.Longsat + 4\beta \quad \text{minutes p.m. GMT}$$

The first part is the time that the Sun passes the meridian at Greenwich (longitude O); the second is the time taken by the Sun to move from the meridian at Greenwich to the meridian at Longsat; the third part is the time taken for the 'shadow' of the satellite to move from Longsat to Longobs. (The Sun appears to move at 15 degrees per hour or one degree in 4 minutes in a westerly direction; the satellite's shadow moves in an easterly direction at about one degree in 20 seconds in the case of the UK).

6. Take Manchester, for example, and the UK broadcast satellite:-

$$A = +53.5 \qquad C = 0.8$$
$$Longobs = -2.2 \qquad D = 0.52$$
$$Longsat = -31 \qquad E = 0.29$$
$$B = +28.8$$

whence $\beta = 2.75$ and $\delta = -7.6^{\circ}$ or $-7^{\circ}36'$, equalled by the Sun on March 2nd 1983 when T_e was 12m 23s (late) and October 13th 1983 when T_e was 13m 31s (early)

Hence T_m = +12.4 + 124 + 11 = 147.4m or 2h 27.4m p.m. on March 2nd
and $\quad\quad$ -13.5 + 124 + 11 = 121.5m or 2h 1.5m p.m. on Oct 13th

7. For other observers in the UK, the following quick approximation may be made to find the change in T_m caused by a change in latitude and longitude from those given above for Manchester :-

$$\Delta T_m = \frac{\Delta Lat}{4} - \frac{\Delta Long}{3} \quad \text{minutes earlier}$$

'SUN-BLINDING'

Approximate intensity and duration of noise-burst at Sun-Satellite conjunction.

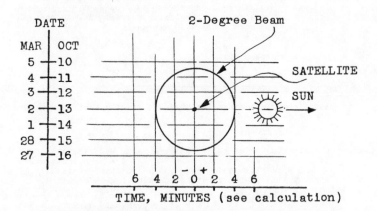

ILLUSTRATING MOTION OF SUN NEAR CONJUNCTION

Antenna diameter, m	0.45	0.65	0.9	1.3	1.9	2.8
Axial gain, dB	32	35	38	41	44	47
Beam width, degrees	4.0	2.8	2.0	1.4	1.0	0.7
Max. extra noise temp. K	300	600	1200	2400	4500	8000
No. minutes burst>$\frac{1}{2}$ pwr.	17	12	9	6	5	4
Adjacent days burst>$\frac{1}{2}$ pwr.	12	8	6	4	3	3

Assuming an effective Sun 0.81 degrees diam. at a temperature of $1.5\ 10^4$K at 12 GHz and a 55% antenna efficiency.

APPENDIX 9

Inter-Modulation Product Generation

Two sinewaves passed together through a non-linear circuit give rise to inter-modulation products. For simplicity a single non-linearity of the form $y = x^2$ will be considered. The two sinewaves are:

$$x = a \sin \alpha t \quad \text{and} \quad x = b \sin \beta t$$

where the angular frequencies of the two waveforms are α and β.

The input to the non-linear circuit $= a \sin \alpha t + b \sin \beta t$

The output $= a^2 \sin^2 \alpha t + b^2 \sin^2 \beta t + ab \sin \alpha t \sin \beta t$

$= 0.5(a^2 + b^2 - a^2 \cos 2\alpha - b^2 \cos 2\beta + ab \cos(\alpha-\beta) - ab \cos(\alpha+\beta))$

This shows that the output waveform consists of a mixture of waveforms, namely:

1. The second harmonic of each input frequency

2. The difference between the two input signal frequencies

3. The sum of the two input signal frequencies

4. A d.c. component depending on the input signal amplitudes

Thus, if WARC'77 channels 4 and 8 were applied to a non-linear amplifier exhibiting square law characteristics, the input frequencies, 11.78502 and 11.36174 GHz would result in the output containing the second harmonic signals, 23.57004 GHz and 23.72348GHz the sum signal 23.64676 GHz and the difference signal 76.72 MHz.

In practice the transfer characteristic of a TWTA is not a square law but something between a straight line linear characteristic and a limiter depending on input level. However, any departure from linear performance causes production of extra components of the general type described. The output power of an amplifier is the sum of powers of wanted and unwanted components. Thus unwanted components not only potentially cause interference problems but diminish the output energy.

CCIR Grading of opinion using a 5 point scale

Five-grade scale	
Quality	Impairment
5 Excellent	5 Imperceptible
4 Good	4 Perceptible, but not annnoying
3 Fair	3 Slightly annoying
2 Poor	2 Annoying
1 Bad	1 Very annoying

APPENDIX 11

THE C - MAC / PACKET SYSTEM

An extract from the Royal Television Society Journal : Television
September/October   1983

Introduction

The subject of technical standards for DBS has been hotly
debated for many years.   There can be few engineers that
have not heard of the battle for video standards but not
many know of the corresponding struggle for a sound system.
Sound was always forgotten, or at least left to last.
However, the sound system proposed is quite revolutionary
and is a major contributor to the cost of the complete
receiver.

At a recent RTS meeting the history of this struggle was
reviewed.   It began with an early decision that today's
technology made it possible to describe a colour picture
by its components, that is luminance (Y) and the colour
difference signals (U and V).

There was also recognition that the only way to transmit
all the sound services that were required was to use
digital coding.

This was the beginning of a struggle that was to last two
years.   It was to be the subject of the most intensive
debate and experiment that the industry has ever seen.

Any historical review is of only passing value because new
developments quickly outdate any record of achievements.
Suffice to say that there was a majority decision by the
European Broadcasting Union this Summer.   The engineers
had done their job and made a recommendation.

It was in effect a statement that if you seek a solution
to a complex problem you would be well advised to tackle
each question separately, that is sequentially, one after
the other.   In engineering terms we call this Time Division
Multiplex (TDM).

298

FIG. 1   TRANSMITTED   SIGNAL   DIAGRAM.

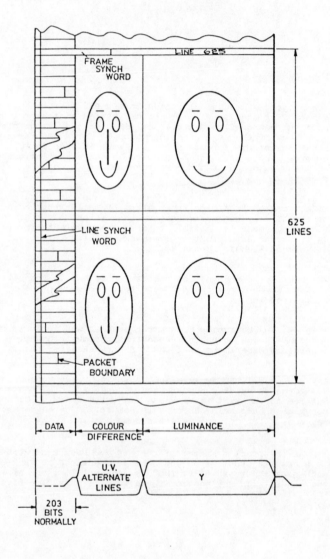

Vision Coding

It was decided that we should transmit the sound (or data)
first, using digital modulation; then the colour components
U and V; followed by the luminance signal Y.   In order to
fit the signals within the horizontal interval of the
television line (64 microseconds) it was necessary to time
compress the video signal before transmission and expand
it later in the receiver.   This is the system we know as
MAC (Multiplexed Analogue Components).   (Figure 1)

Sound: the poor relation

The complete system is best described as "C MAC (Packet)".
The C (Packet) refers to the sound system.

Three sound systems were originally proposed which, for
convenience, were labelled A, B and C.

The Scandinavian countries required four stereo signals to
carry the services of each member country and the only
system capable of this was System 'C'.   This then had to
be the basis of the international standard.

How shall all this data be organised?   Shall we transmit
a 'contents page', telling the receiver where to find the
information (a structure map) or shall we divide the data
into fixed length 'packets' and give each packet an address
to indicate its purpose or destination?

The 'Packet' organisation won the day because, although
complicated, it was extremely flexible and could be expected
to meet the needs of the future whatever it may be.

At this point it is necessary to appreciate that the
receiver that engineers had in mind had little resemblance
to the old receiver with coils and condensors, resistors and
discrete components.   It was designed for the days of the
integrated circuit.   It would appear that using this
technique "complexity has no cost".   Circuit complexity
was only limited by the ideas and imagination of the engineers
concerned.

The proposal was that there would be a large number of
packet addresses; one for each output. The receiver would
have to be told which output was needed and it would then
look up the required packet address. It would then
inspect each packet address as it arrives, select the
right one and redirect the information to the output.

This information will need processing before it can be used
so receiver must first be told what to do with it! There
is, therefore, a need for a number of packets of information
which serve only to guide this intelligent receiver.

There is also a need for packets of information to guide
the viewer! We must make sure the receiver is "user
friendly".

To carry out these two administrative functions we have two
special packet addresses.

The first is address "0" which labels the System Identification
packet (SI).   This is, in effect, a "menu" to tell the
user what services are available.   It can give comprehensive
information on the type of programme and its origin, together
with instructions about which button to press.

The second type of packet is a set of "recipes" for the
receiver to use when preparing the information.   This
will define the bandwidth and prescribe the decoding
system to be used (The Block Interpretation packet, (BI)).

The digital signal employs a 20.25 MHz clock frequency and in
system 'C' this signal is transmitted as Phase Shift Keying
of the Radio Frequency carrier.

There are normally about 3 Mbits per second of data
which can be used for any purpose defined by the BI packet.
If, for example, we decide to transmit a companded stereo
signal we need about 650 Kbits and we have room for 4 services
of this character together with the necessary administrative
overhead of the SI and BI packets.

Alternatively we can transmit the ultimate in sound quality
using linear coding and protect the signal so that it can
be received in conditions of very poor signal strength.
In this case only four such services could be transmitted.

The control mechanism and the signal itself must be protected
from noise and interference.   Again there are options
as to the degree of security provided by the coding system.

There remains the question of how does such a complicated
receiver get started.   What happens when you switch on
or change channel?   How can it operate before it has received
the essential packets of information?   The answer lies in
the synchronising words and the data in Line 625.   This is
the only information outside the packet structure.   It is
transmitted in every frame and contains enough information
to launch the system.

The receiver built to this specification should be able to
receive signals from all sources regardless of their system
choice providing it is within the range offered by this standard.
It permits development of the picture, sound and data
services in a way which is without parallel and should be
satisfactory throughout the eighties, and beyond, in spite
of the speed of technological change.

The receiver is a close relative to the home computer and
has the same potential for the future.   (It will have
within it at least one microprocessor with the associated
data store).

Although the system shows great promise all the problems
have not been solved.

Some Vital Statistics of the C MAC System

A.   Vision Specification

1. Luminance compression 3:2

2. Colour difference signal compression 3:1

3. Transmitted signal bandwidth nominally 9 MHz.

4. Line sync word 7 bits.

5. Frame sync word 64 bits.

B.   Sound/Data Specification

1. Clock frequency 20.25 MHz

2. Normally there are 203 bits per TV line.

3. Packet size is constant at 751 bits.

4. Packet structure is not synchronous (about 1 every 3 lines, that is 162 packet boundaries per frame).

5. The packet contains 91 useful bytes describing up to 64 sound samples.

6. The packet also contains:

   a. 10 bit address
   b. 11 bit error protection
   c. 2 bit continuity/linking signal

7. There are about 24 SI packets/second

8. There are about 12 BI packets/second

9. Line 625 has a total of 1296 bits.

Convention Relating to the Distribution of Programme-Carrying Signals Transmitted by Satellite (Brussels, 21 May 1974)

The Contracting States,

Aware that the use of satellites for the distribution of programme-carrying signals is rapidly growing both in volume and geographical coverage;

Concerned that there is no world-wide system to prevent distributors from distributing programme-carrying signals transmitted by satellite which were not intended for those distributors, and that this lack is likely to hamper the use of satellite communications;

Recognising, in this respect, the importance of the interests of authors, performers, producers of phonograms and broadcasting organisations;

Convinced that an international system should be established under which measures would be provided to prevent distributors from distributing programme-carrying signals transmitted by satellite which were not intended for those distributors;

Conscious of the need not to impair in any way international agreements already in force, including the International Telecommunication Convention and the Radio Regulations annexed to that Convention, and in particular in no way to prejudice wider acceptance of the Rome Convention of October 26, 1961, which affords protection to performers, producers of phonograms and broadcasting organisations,

Have agreed as follows:

ARTICLE 1

For the purposes of this Convention:

(i)   'signal' is an electronically-generated carrier capable of transmitting programmes;

(ii)   'programme' is a body of live or recorded material consisting of images, sounds or both, embodied in signals emitted for the purpose of ultimate distribution;

(iii)   'satellite' is any device in extraterrestrial space capable of transmitting signals;

(iv)   'emitted signal' or 'signal emitted' is any programme-carrying signal that goes to or passes through a satellite;

(v)   'derived signal' is a signal obtained by modifying the technical characteristics of the emitted signal, whether or not there have been one or more intervening fixations;

(vi)   'originating organisation' is the person or legal entity that decides what programme the emitted signals will carry;

(vii)   'distributor' is the person or legal entity that decides that the transmission of the derived signals to the general public or any section thereof should take place;

(viii)   'distribution' is the operation by which a distributor transmits derived signals to the general public or any section thereof.

ARTICLE 2

(1) Each Contracting State undertakes to take adequate measures to prevent the distribution on or from its territory of any programme-carrying signal by any distributor for whom the signal emitted to or passing through the satellite is not intended. This

obligation shall apply where the originating organisation is a national of another Contracting State and where the signal distributed is a derived signal.

(2) In any Contracting State in which the application of the measures referred to in paragraph (1) is limited in time, the duration thereof shall be fixed by its domestic law. The Secretary-General of the United Nations shall be notified in writing of such duration at the time of ratification, acceptance or accession, or if the domestic law comes into force or is changed thereafter, within six months of the coming into force of that law or of its modification.

(3) The obligation provided for in paragraph (1) shall not apply to the distribution of derived signals taken from signals which have already been distributed by a distributor for whom the emitted signals were intended.

ARTICLE 3

This Convention shall not apply where the signals emitted by or on behalf of the originating organisation are intended for direct reception from the satellite by the general public.

ARTICLE 4

No Contracting State shall be required to apply the measures referred to in Article 2(1) where the signal distributed on its territory by a distributor for whom the emitted signal is not intended

  (i) carries short excerpts of the programme carried by the emitted signal, consisting of reports of current events, but only to the extent justified by the informatory purpose of such excerpts, or
 (ii) carries, as quotations, short excerpts of the programme carried by the emitted signal, provided that such quotations are compatible with fair practice and are justified by the informatory purpose of such quotations, or
(iii) carries, where the said territory is that of a Contracting State regarded as a developing country in conformity with the established practice of the General Assembly of the United Nations, a programme carried by the emitted signal, provided that the distribution is solely for the purpose of teaching, including teaching in the framework of adult education, or scientific research.

ARTICLE 5

No Contracting State shall be required to apply this Convention with respect to any signal emitted before this Convention entered into force for that State.

ARTICLE 6

This Convention shall in no way be interpreted to limit or prejudice the protection secured to authors, performers, producers of phonograms, or broadcasting organisations, under any domestic law or international agreement.

ARTICLE 7

This Convention shall in no way be interpreted as limiting the right of any Contracting State to apply its domestic law in order to prevent abuses of monopoly.

ARTICLE 8

(1) Subject to paragraphs (2) and (3), no reservation to this Convention shall be permitted.

(2) Any Contracting State whose domestic law, on May 21, 1974, so provides may, by a written notification deposited with the Secretary-General of the United Nations, declare that, for its purposes, the words 'where the originating organisation is a national of another Contracting State' appearing in Article 2(1) shall be considered as if they were replaced by the words 'where the signal is emitted from the territory of another Contracting State.'

(3) (a) Any Contracting State which, on May 21, 1974, limits or denies protection

with respect to the distribution of programme-carrying signals by means of wires, cable
or other similar communications channels to subscribing members of the public may,
by a written notification deposited with the Secretary-General of the United Nations,
declare that, to the extent that and as long as its domestic law limits or denies protection,
it will not apply this Convention to such distributions.

(b) Any State that has deposited a notification in accordance with subparagraph (a)
shall notify the Secretary-General of the United Nations in writing, within six months
of their coming into force, of any changes in its domestic law whereby the reservation
under that subparagraph becomes inapplicable or more limited in scope.

ARTICLE 9

(1) This Convention shall be deposited with the Secretary-General of the United
Nations. It shall be open until March 31, 1975, for signature by any State that is a
member of the United Nations, any of the Specialised Agencies brought into relationship
with the United Nations, or the International Atomic Energy Agency, or is a party to
the Statute of the International Court of Justice.

(2) This Convention shall be subject to ratification or acceptance by the signatory
States. It shall be open for accession by any State referred to in paragraph (1).

(3) Instruments of ratification, acceptance or accession shall be deposited with the
Secretary-General of the United Nations.

(4) It is understood that, at the time a State becomes bound by this Convention, it will
be in a position in accordance with its domestic law to give effect to the provisions of the
Convention.

ARTICLE 10

(1) This Convention shall enter into force three months after the deposit of the fifth
instrument of ratification, acceptance or accession.

(2) For each State ratifying, accepting or acceding to this Convention after the deposit
of the fifth instrument of ratification, acceptance or accession, this Convention shall
enter into force three months after the deposit of its instrument.

ARTICLE 11

(1) Any Contracting State may denounce this Convention by written notification
deposited with the Secretary-General of the United Nations.

(2) Denunciation shall take effect twelve months after the date on which the notification
referred to in paragraph (1) is received.

ARTICLE 12

(1) This Convention shall be signed in a single copy in English, French, Russian and
Spanish, the four texts being equally authentic.

(2) Official texts shall be established by the Director-General of the United Nations
Educational, Scientific and Cultural Organisation and the Director-General of the
World Intellectual Property Organisation, after consultation with the interested
Governments, in the Arabic, Dutch, German, Italian and Portuguese languages.

(3) The Secretary-General of the United Nations shall notify the States referred to in
Article 9(1), as well as the Director-General of the United Nations Educational, Scientific and Cultural Organisation, the Director-General of the World Intellectual Property Organisation, the Director-General of the International Labour Office and the
Secretary-General of the International Telecommunication Union, of

(i) signatures to this Convention;
(ii) the deposit of instruments of ratification, acceptance or accession;
(iii) the date of entry into force of this Convention under Article 10(1);
(iv) the deposit of any notification relating to Article 2(2) or Article 8(2) or (3),
together with its text;
(v) the receipt of notifications of denunciation.

(4) The Secretary-General of the United Nations shall transmit two certified copies of
this Convention to all States referred to in Article 9(1).

TREATY ON PRINCIPLES GOVERNING THE ACTIVITIES OF STATES
IN THE EXPLORATION AND USE OF OUTER SPACE, INCLUDING
THE MOON AND OTHER CELESTIAL BODIES
(OUTER SPACE TREATY), 1967

The States Parties to this Treaty,
Inspired by the great prospects opening up before mankind as a result of man's entry
into outer space,
   Recognizing the common interest of all mankind in the progress of the exploration
and use of outer space for peaceful purposes,
   Believing that the exploration and use of outer space should be carried on for the
benefit of all peoples irrespective of the degree of their economic or scientific
development,
   Desiring to contribute to broad international co-operation in the scientific as well
as the legal aspects of the exploration and use of outer space for peaceful purposes,
   Believing that such co-operation will contribute to the development of mutual
understanding and to the strengthening of friendly relations between States and
peoples,
   Recalling resolution 1962 (XVIII), entitled "Declaration of Legal Principles Governing the Activities of States in the Exploration and Use of Outer Space," which was
adopted unanimously by the United Nations General Assembly on 13 December
1963,
   Recalling resolution 1884 (XVIII), calling upon States to refrain from placing in orbit around the Earth any objects carrying nuclear weapons or any other kinds of
weapons of mass destruction or from installing such weapons on celestial bodies,
which was adopted unanimously by the United Nations General Assembly on 17
October 1963,
   Taking account of United Nations General Assembly resolution 110 (11) of 3
November 1947, which condemned propaganda designed or likely to provoke or encourage any threat to the peace, breach of the peace or act of aggression, and considering that the aforementioned resolution is applicable to outer space,
   Convinced that a Treaty on Principles Governing the Activities of States in the Exploration and Use of Outer Space, including the Moon and Other Celestial Bodies,
will further the purposes and principles of the Charter of the United Nations,
   Have agreed on the following:

ARTICLE I

The exploration and use of outer space, including the Moon and other celestial
bodies, shall be carried out for the benefit and in the interests of all countries, irrespective of their degree of economic or scientific development, and shall be the
province of all mankind.
   Outer space, including the Moon and other celestial bodies, shall be free, for exploration and use by all States without discrimination of any kind, on a basis of
equality and in accordance with international law, and there shall be free access to
all areas of celestial bodies.
   There shall be freedom of scientific investigation in outer space, including the
Moon and other celestial bodies, and States shall facilitate and encourage international co-operation in such investigation.

ARTICLE II

Outer space, including the Moon and other celestial bodies, is not subject to national appropriation by claim of sovereignty, by means of use or occupation, or by any other means.

ARTICLE III

States Parties to the Treaty shall carry on activities in the exploration and use of outer space, including the Moon and other celestial bodies, in accordance with international law, including the Charter of the United Nations, in the interest of maintaining peace and security and promoting international co-operation and understanding.

ARTICLE IV

States Parties to the Treaty undertake not to place in orbit around the Earth any objects carrying nuclear weapons or any other kinds of weapons of mass destruction, install such weapons on celestial bodies, or station weapons in outer space in any other manner.

The Moon and other celestial bodies shall be used by all States Parties to the Treaty exclusively for peaceful purposes. The establishment of military bases, installations and fortifications, the testing of any type of weapons and the conduct of military manoeuvres on celestial bodies shall be forbidden. The use of military personnel for scientific research or for any other peaceful purposes shall not be prohibited. The use of any equipment or facility necessary for peaceful exploration of the Moon and other celestial bodies shall also not be prohibited

ARTICLE V

States Parties to the Treaty shall regard astronauts as envoys of mankind in outer space and shall render to them all possible assistance in the event of accident, distress, or emergency landing on the territory of another State Party or on the high seas. When astronauts make such a landing, they shall be safely and promptly returned to the State of registry of their space vehicle.

In carrying on activities in outer space and on celestial bodies, the astronauts of one State Party shall render all possible assistance to the astronauts of other States Parties.

States Parties to the Treaty shall immediately inform the other States Parties to the Treaty or the Secretary-General of the United Nations of any phenomena they discover in outer space, including the Moon and other celestial bodies, which could constitute a danger to the life or health of astronauts.

ARTICLE VI

States Parties to the Treaty shall bear international responsibility for national activities in outer space, including the Moon and other celestial bodies, whether such activities are carried on by governmental agencies or by non-governmental entities, and for assuring that national activities are carried out in conformity with the provisions set forth in the present Treaty. The activities of non-governmental entities in outer space, including the Moon and other celestial bodies, shall require authorization and continuing supervision by the appropriate State Party to the Treaty. When activities are carried on in outer space, including the Moon and other celestial bodies, by an international organization, responsibility for compliance with this Treaty shall be borne both by the international organization and by the States Parties to the Treaty participating in such organization.

ARTICLE VII

Each State Party to the Treaty that launches or procures the launching of an object into outer space, including the Moon and other celestial bodies, and each State Party from whose territory or facility an object is launched, is internationally liable for damage to another State Party to the Treaty or to its natural or juridical persons by such object or its component parts on the Earth, in air space or in outer space, including the Moon and other celestial bodies.

ARTICLE VIII

A State Party to the Treaty on whose registry an object launched into outer space is carried shall retain jurisdiction and control over such object, and over any personnel thereof, while in outer space or on a celestial body. Ownership of objects launched into outer space, including objects landed or constructed on a celestial body, and of their component parts, is not affected by their presence in outer space or on a celestial body or by their return to the Earth. Such objects or component parts found beyond the limits of the State Party to the Treaty on whose registry they are carried shall be returned to that State Party, which shall, upon request, furnish identifying data prior to their return.

ARTICLE IX

In the exploration and use of outer space, including the Moon and other celestial bodies, States Parties to the Treaty shall be guided by the principle of co-operation and mutual assistance and shall conduct all their activities in outer space, including the Moon and other celestial bodies, with due regard to the corresponding interests of all other States Parties to the Treaty. States Parties to the Treaty shall pursue studies of outer space, including the Moon and other celestial bodies, and conduct exploration of them so as to avoid their harmful contamination and also adverse changes in the environment of the Earth resulting from the introduction of extraterrestrial matter and, where necessary, shall adopt appropriate measures for this purpose. If a State Party to the Treaty has reason to believe that an activity or experiment planned by it or its nationals in outer space, including the Moon and other celestial bodies, would cause potentially harmful interference with activities of other States Parties in the peaceful exploration and use of outer space, including the Moon and other celestial bodies, it shall undertake appropriate international consultations before proceeding with any such activity or experiment. A State Party to the Treaty which has reason to believe that an activity or experiment planned by another State Party in outer space, including the Moon and other celestial bodies, would cause potentially harmful interference with activities in the peaceful exploration and use of outer space, including the Moon and other celestial bodies, may request consultation concerning the activity or experiment.

ARTICLE X

In order to promote international co-operation in the exploration and use of outer space, including the Moon and other celestial bodies, in conformity with the purposes of this Treaty, the States Parties to the Treaty shall consider on a basis of equality any requests by other States Parties to the Treaty to be afforded an opportunity to observe the flight of space objects launched by those States.

The nature of such an opportunity for observation and the conditions under which it could be afforded shall be determined by agreement between the States concerned.

ARTICLE XI

In order to promote international co-operation in the peaceful exploration and use of outer space, States Parties to the Treaty conducting activities in outer space, including the Moon and other celestial bodies, agree to inform the Secretary-General of the United Nations as well as the public and the international scientific community, to the greatest extent feasible and practicable, of the nature, conduct, locations and results of such activities. On receiving the said information, the Secretary-General of the United Nations should be prepared to disseminate it immediately and effectively.

ARTICLE XII

All stations, installations, equipment and space vehicles on the Moon and other celestial bodies shall be open to representatives of other States Parties to the Treaty on a basis of reciprocity. Such representatives shall give reasonable advance notice of a projected visit, in order that appropriate consultations may be held and that maximum precautions may be taken to assure safety and to avoid interference with normal operations in the facility to be visited.

ARTICLE XIII

The provisions of this Treaty shall apply to the activities of States Parties to the Treaty in the exploration and use of outer space, including the Moon and other celestial bodies, whether such activities are carried on by a single State Party to the Treaty or jointly with other States, including cases where they are carried on within the framework of international intergovernmental organizations.

Any practical questions arising in connexion with activities carried on by international intergovernmental organizations in the exploration and use of outer space, including the Moon and other celestial bodies, shall be resolved by the States Parties to the Treaty either with the appropriate international organization or with one or more States members of that international organization, which are Parties to this Treaty.

ARTICLE XIV

1. This Treaty shall be open to all States for signature. Any State which does not sign this Treaty before its entry into force in accordance with paragraph 3 of this article may accede to it at any time.

2. This Treaty shall be subject to ratification by signatory States. Instruments of ratification and instruments of accession shall be deposited with the governments of the Union of Soviet Socialist Republics, the United Kingdom of Great Britain and Northern Ireland and the United States of America, which are hereby designated the Depository Governments.

3. This Treaty shall enter into force upon the deposit of instruments of ratification by five Governments including the Governments designated as Depository Governments under this Treaty.

4. For States whose instruments of ratification or accession are deposited subsequent to the entry into force of this Treaty, it shall enter into force on the date of the deposit of their instruments of ratification or accession.

5. The Depository Governments shall promptly inform all signatory and acceding States of the date of each signature, the date of deposit of each instrument of ratification of and accession to this Treaty, the date of its entry into force and other notices.

6. This Treaty shall be registered by the Depository Governments pursuant to Article 102 of the Charter of the United Nations.

ARTICLE XV

Any State Party to the Treaty may propose amendments to this Treaty. Amendments shall enter into force for each State Party to the Treaty accepting the amendments upon their acceptance by a majority of the States Parties to the Treaty and thereafter for each remaining State Party to the Treaty on the date of acceptance by it.

ARTICLE XVI

Any State Party to the Treaty may give notice of its withdrawal from the Treaty one year after its entry into force by written notification to the Depository Governments. Such withdrawal shall take effect one year from the date of receipt of this notification.

ARTICLE XVII

This Treaty, of which the Chinese, English, French, Russian and Spanish texts are equally authentic, shall be deposited in the archives of the Depository Governments. Duly certified copies of this Treaty shall be transmitted by the Depository Governments to the Governments of the signatory and acceding States.

In witness whereof the undersigned, duly authorized, have signed this Treaty.

Done in triplicate at the cities of London, Moscow and Washington, this twenty-seventh day of January, one thousand nine hundred sixty-seven.

Appendix 14

Annex 1
SPG 2147 rev
Feb. 83

**UNITED
NATIONS**

 General Assembly

Distr.
GENERAL

A/RES/37/92
4 February 1983

Thirty-seventh session
Agenda items 62, 63 and 131

RESOLUTION ADOPTED BY THE GENERAL ASSEMBLY

[on the report of the Special Political Committee (A/37/646)]

37/92. Principles Governing the Use by States of Artificial Earth
Satellites for International Direct Television Broadcasting

The General Assembly,

Recalling its resolution 2916 (XXVII) of 9 November 1972, in which it stressed
the necessity of elaborating principles governing the use by States of artificial
earth satellites for international direct television broadcasting, and mindful of
the importance of concluding an international agreement or agreements,

Recalling further its resolutions 3182 (XXVIII) of 18 December 1973,
3234 (XXIX) of 12 November 1974, 3388 (XXX) of 18 November 1975, 31/8 of
8 November 1976, 32/196 of 20 December 1977, 33/16 of 10 November 1978, 34/66 of
5 December 1979 and 35/14 of 3 November 1980, and its resolution 36/35 of
18 November 1981 in which it decided to consider at its thirty-seventh session the
adoption of a draft set of principles governing the use by States of artificial
earth satellites for international direct television broadcasting,

Noting with appreciation the efforts made in the Committee on the Peaceful
Uses of Outer Space and its Legal Sub-Committee to comply with the directives
issued in the above-mentioned resolutions,

Considering that several experiments of direct broadcasting by satellite have
been carried out and that a number of direct broadcasting satellite systems are
operational in some countries and may be commercialized in the very near future,

Taking into consideration that the operation of international direct
broadcasting satellites will have significant international political, economic,
social and cultural implications,

83-02774 0933Z (E)   /...

A/RES/37/92
Page 2

Believing that the establishment of principles for international direct television broadcasting will contribute to the strengthening of international co-operation in this field and further the purposes and principles of the Charter of the United Nations,

Adopts the Principles Governing the Use by States of Artificial Earth Satellites for International Direct Television Broadcasting set forth in the annex to the present resolution.

100th plenary meeting
10 December 1982

/...

ANNEX

Principles Governing the Use by States of Artificial Earth Satellites
for International Direct Television Broadcasting

A. Purposes and objectives

1. Activities in the field of international direct television broadcasting
by satellite should be carried out in a manner compatible with the sovereign rights
of States, including the principle of non-intervention, as well as with the right
of everyone to seek, receive and impart information and ideas as enshrined in the
relevant United Nations instruments.

2. Such activities should promote the free dissemination and mutual exchange
of information and knowledge in cultural and scientific fields, assist in
educational, social and economic development, particularly in the developing
countries, enhance the qualities of life of all peoples and provide recreation with
due respect to the political and cultural integrity of States.

3. These activities should accordingly be carried out in a manner compatible
with the development of mutual understanding and the strengthening of friendly
relations and co-operation among all States and peoples in the interest of
maintaining international peace and security.

B. Applicability of international law

4. Activities in the field of international direct television broadcasting
by satellite should be conducted in accordance with international law, including
the Charter of the United Nations, the Treaty on Principles Governing the
Activities of States in the Exploration and Use of Outer Space, including the Moon
and Other Celestial Bodies, of 27 January 1967, the relevant provisions of the
International Telecommunication Convention and its Radio Regulations and of
International instruments relating to friendly relations and co-operation among
States and to human rights.

C. Rights and benefits

5. Every State has an equal right to conduct activities in the field of
international direct television broadcasting by satellite and to authorize such
activities by persons and entities under its jurisdiction. All States and peoples
are entitled to and should enjoy the benefits from such activities. Access to the
technology in this field should be available to all States without discrimination
on terms mutually agreed by all concerned.

/...

D. International co-operation

6. Activities in the field of international direct television broadcasting by satellite should be based upon and encourage international co-operation. Such co-operation should be the subject of appropriate arrangements. Special consideration should be given to the needs of the developing countries in the use of international direct television broadcasting by satellite for the purpose of accelerating their national development.

E. Peaceful settlement of disputes

7. Any international dispute that may arise from activities covered by these principles should be settled through established procedures for the peaceful settlement of disputes agreed upon by the parties to the dispute in accordance with the provisions of the Charter of the United Nations.

F. State responsibility

8. States should bear international responsibility for activities in the field of international direct television broadcasting by satellite carried out by them or under their jurisdiction and for the conformity of any such activities with the principles set forth in this document.

9. When international direct television broadcasting by satellite is carried out by an international intergovernmental organization, the responsibility referred to in paragraph 8 above should be borne both by that organization and by the States participating in it.

G. Duty and right to consult

10. Any broadcasting or receiving State within an international direct television broadcasting satellite service established between them requested to do so by any other broadcasting or receiving State within the same service should promptly enter into consultations with the requesting State regarding its activities in the field of international direct television broadcasting by satellite, without prejudice to other consultations which these States may undertake with any other State on that subject.

H. Copyright and neighbouring rights

11. Without prejudice to the relevant provisions of international law, States should co-operate on a bilateral and multilateral basis for protection of copyright and neighbouring rights by means of appropriate agreements between the interested States or the competent legal entities acting under their jurisdiction. In such co-operation they should give special consideration to the interests of developing countries in the use of direct television broadcasting for the purpose of accelerating their national development.

/...

A/RES/37/92
Page 5

I. Notification to the United Nations

12. In order to promote international co-operation in the peaceful exploration and use of outer space, States conducting or authorizing activities in the field of international direct television broadcasting by satellite should inform the Secretary-General of the United Nations, to the greatest extent possible, of the nature of such activities. On receiving this information, the Secretary-General should disseminate it immediately and effectively to the relevant specialized agencies, as well as to the public and the international scientific community.

J. Consultations and agreements between States

13. A State which intends to establish or authorize the establishment of an international direct television broadcasting satellite service shall without delay notify the proposed receiving State or States of such intention and shall promptly enter into consultation with any of those States which so requests.

14. An international direct television broadcasting satellite service shall only be established after the conditions set forth in paragraph 13 above have been met and on the basis of agreements and/or arrangements in conformity with the relevant instruments of the International Telecommunication Union and in accordance with these principles.

15. With respect to the unavoidable overspill of the radiation of the satellite signal, the relevant instruments of the International Telecommunication Union shall be exclusively applicable.

/...

ANNEX 2
SPG 2147 dt rev.
Feb. 83
(n'existe qu'en anglais)

Thirty-seventh General Assembly
100th Plenary Meeting (PM)

Press Release GA/6769
10 December 1982

ASSEMBLY PLENARY -- ANNEX X

Vote on Direct Television Broadcasting

The Assembly adopted the resolution on draft principles for direct
television broadcasting by satellite (document A/37/646) by a recorded vote of
107 in favour to 13 against, with 13 abstentions, as follows:

In favour: Afghanistan, Algeria, Argentina, Bahrain, Bangladesh,
Barbados, Benin, Bhutan, Bolivia, Botswana, Brazil, Bulgaria, Burma,
Burundi, Byelorussia, Central African Republic, Chad, Chile, China,
Colombia, Comoros, Congo, Cuba, Cyprus, Czechoslovakia, Democratic
Kampuchea, Democratic Yemen, Djibouti, Dominican Republic, Ecuador,
Egypt, El Salvador, Ethiopia, Fiji, Gabon, Gambia, German Democratic
Republic, Ghana, Guyana, Haiti, Honduras, Hungary, India, Indonesia,
Iran, Iraq, Jamaica, Jordan, Kenya, Kuwait, Lao People's Democratic
Republic, Liberia, Libya, Madagascar, Malaysia, Maldives, Mali, Malta,
Mauritania, Mauritius, Mexico, Mongolia, Mozambique, Nepal, Nicaragua,
Niger, Nigeria, Oman, Pakistan, Panama, Papua New Guinea, Peru,
Philippines, Poland, Qatar, Romania, Rwanda, Sao Tome and Principe, Saudi
Arabia, Senegal, Sierra Leone, Singapore, Solomon Islands, Somalia, Sri
Lanka, Sudan, Suriname, Syria, Thailand, Togo, Trinidad and Tobago,
Tunisia, Turkey, Uganda, Ukraine, USSR, United Arab Emirates, United
Republic of Cameroon, United Republic of Tanzania, Upper Volta, Uruguay,
Venezuela, Viet Nam, Yemen, Yugoslavia, Zaire, Zambia.

Against: Belgium, Denmark, Federal Republic of Germany, Iceland, Israel,
Italy, Japan, Luxembourg, Netherlands, Norway, Spain, United Kingdom,
United States.

Abstaining: Australia, Austria, Canada, Finland, France, Greece,
Ireland, Lebanon, Malawi, Morocco, New Zealand, Portugal, Sweden.

Absent: Albania, Angola, Antigua and Barbuda, Bahamas, Belize, Cape
Verde, Costa Rica, Dominica, Equatorial Guinea, Grenada, Guatemala,
Guinea, Guinea-Bissau, Ivory Coast, Lesotho, Paraguay, Saint Lucia, Saint
Vincent, Samoa, Seychelles, Swaziland, Vanuatu, Zimbabwe.

(END OF ANNEX X)

Energy-dispersal

Although satellite broadcasting has preferential treatment when it comes to sharing the down-link band, 11.7 – 12.5 GHz, this is not so in the band used for the up-link or feeder link to the broadcast satellites (probably 17.3 – 18.1 GHz). There may be other satellites sharing the latter band that are susceptible to interference and may deserve some special protection. This can arise, for example, when a television transmission is of a blank screen or a caption on a uniform background, so that the r.f. spectrum consits of a few discrete components of fairly high power. Such a discrete component may, if it falls on one particular channel of a narrow-band digital telephony service carried by a neighbouring communication satellite, cause sufficient interference to make the channel fail completely, until the television picture becomes 'busy' again.

To avoid interference from television to narrow-band communications, a dispersal waveform has been proposed which, when added to the video signal before transmission, ensures that any discrete lines in the television spectrum are 'blurred out' over many telephone channels. The proposed waveform is triangular with a frequency deviation of 600 kHz and a period of a multiple of the field period. The resulting dispersed carrier power is about 22 dB lower than if it all fell in one 4 kHz-wide channel. This reduction in power density may not be as effective as the figure suggests for reasons connected with the error-protection system employed by the digital channel. If the linearly-dispersed television component spends too long sweeping through the 4 kKz digital channel, the interference may affect not one or two, but several adjacent bits, amounting to an error burst and overloading the system. A higher dispersal frequency would be better in this respect but would probably be more difficult to remove in the receiver.

The dispersal waveform would not need to be removed in the receiver if it could be removed in the satellite, before transmission on the down-link. However, it is different and complex to remove the waveform in the satellite without first demodulating, and subsequently remodulating – a complication that the satellite transponder designers would rather not contemplate. It is likely that satellite transponders will therefore be 'transparent'; what goes up will come down simply transposed in frequency, and the receiver will have to remove the dispersal waveform with as little residual degradation as possible.

Additional dispersal will arise from the process of 'scrambling' the vision signal in order to provide conditional access, as described in Chapter 5.7.3.

WARC 1977 ALLOCATIONS - PRINTED IN ORBIT SLOT ORDER

Country	Channels Allocated	Orbit Posn. (degrees)	Boresight Lat. Long. (degrees)		Beamwidth to 3 dB points (degrees)		Centre EIRP (dBW)	Approx. R.F. power (watts)
Oceania	4,8,12,16	-160	-16.3	-145	4.34	3.54	63.5	2491
Andorra	4,8,12,16,20	-37	42.5	1.6	.6	.6	61.5	36
Vatican	27,31,35,39	-37	41.8	12.4	.6	.6	65.2	86
Vatican	23	-37	41.5	10.8	2	.6	63.6	199
Gambia	3,7,11,15,19	-37	13.4	-15.1	.79	.6	63.3	73
Guinea	1,5,9,13,17	-37	10.2	-11	1.58	1.04	63.4	260
Liechtenstein	3,7,11,15,19	-37	47.1	9.5	.6	.6	62.4	45.
Monaco	21,25,29,33,37	-37	43.7	7.4	.6	.6	62.4	45
Mali	2,6,10,14,18	-37	19	-2	2.66	1.26	63.2	507
Mali	4,8,12,16,20	-37	13.2	-7.6	1.74	1.24	63.7	366
Mauritania	22,26,30,34,38	-37	18.5	-12.2	2.62	1.87	62.8	676
Mauritania	24,28,32,36,40	-37	23.4	-7.8	1.63	1.1	63	259
Senegal	21,25,29,33,37	-37	13.8	-14.4	1.46	1.04	63.6	251
San Marino	1,5,9,13,17	-37	43.7	12.6	.6	.6	62.4	45
Azores	3,7,11,15,19	-31	36.1	-23.4	2.56	.7	63	259
Canary Is.	23,27,31,35,39	-31	28.4	-15.7	1.51	.6	62.8	127
Cape Verde Is.	4,8,12,16,20	-31	16	-24	.86	.7	62.2	72
Ivory Coast	22,26,30,34,38	-31	7.5	-5.6	1.6	1.22	63.7	331
Spain	23,27,31,35,39	-31	39.9	-3.1	2.1	1.14	63.9	425
United Kingdom	4,8,12,16,20	-31	53.8	-3.5	1.84	.72	65	303
Guinea-Bissau	2,6,10,14,18	-31	12	-15	.9	.6	63.1	79
Upper Volta	21,25,29,33,37	-31	12.2	-1.5	1.45	1.14	64	300
Ireland	2,6,10,14,18	-31	53.2	-8.2	.84	.6	64.2	96
Iceland	21,25,29,33,37	-31	64.9	-19	1	.6	65.8	165
Liberia	3,7,11,15	-31	6.6	-9.3	1.22	.7	63.2	129
Portugal	3,7,11,15,19	-31	39.6	-8	.92	.6	63.4	87
Sierra Leone	23,27,31,35,39	-31	8.6	-11.8	.78	.68	63.4	84
Algeria	2,6,10,14,18	-25	33.2	4.2	2.45	1.25	63.4	485
Algeria	4,8,12,16,20	-25	25.5	1.6	3.64	2.16	62.8	1085
Ghana	23,27,31,35,39	-25	7.9	-1.2	1.48	1.06	63.6	260
Libya	1,5,9,13,17	-25	26	21.4	2.5	1.04	63.5	421
Libya	3,7,11,15,19	-25	27.2	13.1	2.36	1.12	63	382
Morocco	21,25,29,33,37	-25	29.2	-9	2.72	1.47	63.3	619
Niger	24,28,32,36,40	-25	16.8	8.3	2.54	2.08	64.5	1078
Togo	2,6,10,14,18	-25	8.6	.8	1.52	.6	63.4	144
Tunisia	22,26,30,34	-25	33.5	9.5	1.88	.72	63.8	235
Tunisia	38	-25	32	2.5	3.59	1.75	61.9	704
Austria	4,8,12,16,20	-19	47.5	12.2	1.14	.63	64.1	133
Belgium	21,25,29,33,37	-19	50.6	4.6	.82	.6	64.2	93
Benin	3,7,11,15,19	-19	9.5	2.2	1.44	68	63.3	151
West Germany	2,6,10,14,18	-19	49.9	9.6	1.62	.72	65.5	299
France	1,5,9,13,17	-19	45.9	2.6	2.5	.98	63.8	425
Eq. Guinea	23,27,31,35,39	-19	1.5	10.3	.68	.6	63.8	70
Netherlands	23,27,31,35,39	-19	52	5.4	.76	.6	64.4	90
Italy	24,28,32,36,40	-19	41.3	12.3	2.38	.98	64.1	434
Luxembourg	3,7,11,15,19	-19	49.8	6	.6	.6	62.9	50

Orbit Slots -

Countries	Channels Allocated	Orbit Posn. (degrees)	Boresight Lat. Long. (degrees)		Beamwidth to 3 dB points (degrees)		Centre EIRP (dBW)	Approx. R.F. power (watts)
Nigeria	22,26,30,34,38	-19	9.4	7.8	2.16	2.02	63.9	775
Namibia	25,29,33,37	-19	-21.6	17.5	2.66	1.9	64.7	1080
Switzerland	22,26,30,34,38	-19	46.6	8.2	.98	.7	64.1	127
Zaire	4,8,12,16,20	-19	0	22.4	2.16	1.88	64.7	868
Zaire	2,6,10,14,18	-19	-6.8	21.3	2.8	1.52	64.6	889
Angola	23,27,31,35,39	-13	-12	16.5	3.09	2.26	64.1	1300
Central Af. Rpb	24,28,32,36,40	-13	6.3	21	2.25	1.68	64.3	737
Cameroun	1,5,9,13,17	-13	6.2	12.7	2.54	1.68	63.4	676
Congo	22,26,30,34,38	-13	-.7	14.6	2.02	1.18	63.8	414
Gabon	3,7,11,15,19	-13	-.6	11.8	1.43	1.12	63.3	248
Israel	25,29,33,37	-13	31.4	34.9	.94	.6	63.8	98
Malta	4,8,12,16	-13	35.9	14.3	.6	.6	61	32
Sao Tome	4,8,12,16,20	-13	.8	7	.6	.6	61.4	35
Chad	2,6,10,14,18	-13	15.5	18.1	3.4	1.72	64	1064
Albania	22,26,30,34,38	-7	41.3	19.8	.68	.6	63.8	70
Egypt	4,8,12,16,20	-7	26.8	29.7	2.33	1.72	63.1	592
Sudan	23,27,31,35,39	-7	7.5	29.2	2.?4	1.12	64.4	522
Sudan	22,26,30,34,38	-7	12.7	28.9	2.26	1.96	63.5	718
Sudan	24,28,32,36,40	-7	19	30.4	2.44	1.52	63.3	574
Yugoslavia	21,25,29,33,37	-7	43.7	18.4	1.68	.66	65.2	265
Yugoslavia	23,27,31,35,39	-7	43.7	18.4	1.68	.66	65.2	265
Botswana	2,6,10,14,18	-1	-22.2	23.3	2.13	1.5	63.7	542
Bulgaria	4,8,12,16,20	-1	43	25	1.04	.6	63.6	103
East Germany	21,25,29,33,37	-1	52.1	12.6	.83	.63	64.2	99
Hungary	22,26,30,34,38	-1	47.2	19.5	.92	.6	64	100
Zimbabwe	22,26,30,34,38	-1	-18.8	29.6	1.46	1.36	64.2	378
Mozambique	4,8,12,16,20	-1	-18	34	3.57	1.38	64.2	938
Malawi	24,28,32,36,40	-1	-13	34.1	1.54	.6	64.2	176
Poland	1,5,9,13,17	-1	51.8	19.3	1.46	.64	64.1	173
Romania	2,6,10,14,18	-1	45.7	25	1.38	.66	63.8	158
Swaziland	1,5,9,13,17	-1	-26.5	31.5	.62	.6	62.8	51
Czechoslovakia	3,7,11,15,19	-1	49.3	17.3	1.47	.6	63.8	153
Zambia	3,7,11,15,19	-1	-13.1	27.5	2.38	1.48	63.7	598
Cyprus	21,25,29,33,37	5	35.1	33.1	.6	.6	63.6	59
Denmark	12,16,20	5	57.1	12.3	1.2	.6	64.3	140
Denmark	24,36	5	61.5	17	2	1	67.5	814
Denmark	27,35	5	61	-19.5	2.2	.8	66.2	531
Finland	2,6,10	5	64.5	22.5	1.38	.76	67.7	447
Finland	22,26	5	61.5	17	2	1	67.7	853
Greece	3,7,11,15,19	5	38.2	24.7	1.78	.98	63.3	270
South Africa	21,25,29,33,37	5	-28	24.5	3.13	1.68	64.1	979
Iceland	23,31,39	5	61	-19.5	2.2	.8	66.3	543
Lesotho	24,28,32,36,40	5	-29.8	27.8	.66	.6	64.2	75
Norway	14,18,38	5	64.1	13.1	1.84	.88	65	370
Norway	28,32	5	61.5	17	2	1	66.8	693
Sweden	4,8,34	5	61	16.2	1.04	.98	67.1	378

Orbit Slots -

Countries	Channels Allocated	Orbit Posn. (degrees)	Boresight Lat. Long. (degrees)		Beamwidth to 3 dB points (degrees)		Centre EIRP (dBW)	Approx. R.F. power (watts)
Sweden	30,40	5	61.5	17	2	1	67.1	743
Turkey	1,5,9,13,17	5	38.9	34.4	2.68	1.04	63.7	473
Burundi	22,26,30,34,38	11	-3.1	29.9	.71	.6	63.4	67
Burundi	22,26,30,34,38	11	-3.1	29.9	.71	.6	63.4	67
Burundi	22,26,30,34,38	11	-3.1	29.9	.71	.6	63.4	67
Iraq	24,28,32,36,40	11	32.8	43.6	1.88	.96	63.3	279
Jordan	23,27,31,35,39	11	31.4	35.8	.84	.78	63.1	96
Kenya	21,25,29,33,37	11	1.1	37.9	2.29	1.56	63.7	606
Lebanon	3,7,11,15,19	11	33.9	35.8	.6	.6	61.6	37
Rwanda	4,8,12,16,20	11	-2.1	30	.66	.6	64.8	86
Syria	22,26,30,34	11	34.9	38.3	1.04	.9	63.2	141
Syria	38	11	34.2	37.6	1.32	.88	63.4	184
Tanzania	23,27,31,35,39	11	-6.2	34.6	2.41	1.72	63.7	703
Uganda	3,7,11,15,19	11	1.2	32.3	1.46	1.12	63.2	247
Yemen	2,6,10,14,18	11	15.1	44.3	1.14	.7	62.6	105
South Yemen	1,5,9,13,17	11	15.2	48.8	1.76	1.54	62.8	374
Saudi Arabia	4,8,12,16,20	17	23.8	41.1	3.52	1.68	62.7	797
Saudi Arabia	2,6,10,14,18	17	24.6	48.3	3.84	1.2	62.7	621
Saudi Arabia	23,35,39	17	24.8	52.3	2.68	.7	63.2	283
Bahrain	27,31,35,39	17	26.1	50.5	.6	.6	60.8	31
Kuwait	22,26,30,34,38	17	29.2	47.6	.68	.6	63.1	60
Oman	24,28,32,36,40	17	21	55.6	1.88	1.02	63.3	297
Qatar	1,5,9,13,17	17	25.3	51.1	.6	.6	61.8	39
U. A. Emirates	21,25,29,33,37	17	24.2	53.6	.98	.8	63.2	118
Afors & Issas	21,25,29,33,37	23	11.6	42.5	.6	.6	62.5	46
Byelorussia	21,25	23	52.6	27.8	1.08	.72	64.8	170
Ethiopia	22,26,30,34,38	23	9.1	39.7	3.5	2.4	63.4	1331
Somalia	3,7,11,15,19	23	6.4	45	3.26	1.54	62.3	617
Ukraine	29,33,37	23	48.4	31.2	2.32	.96	64.6	465
U.S.S.R.	27,31,35,39	23	47	36	3.7	1.43	65.2	1269
U.S.S.R.	4,8,12,16	23	57.4	41.5	3.08	1.56	66.7	1628
U.S.S.R.	3,7,11,15,19,23	23	56.6	24.7	.88	.64	65	129
U.S.S.R.	1,5,9,13,17,23	23	40.8	45.6	2.16	.6	63.9	230
U.S.S.R.	20	23	63.1	32.4	1.18	.6	66.6	234
Comoro Is.	3,7,11,15	29	-12.1	44.1	.76	.6	63.1	67
Mauritius	2,6,10,14,18	29	-18.9	59.8	1.62	1.24	64	365
Mauritius	4,8,12,16	29	-13.9	56.8	1.56	1.38	63.7	365
Malagasy	1,5,9,13,17	29	-18.8	46.6	2.72	1.14	63.3	480
Mayotte	24,28,32,36,40	29	-12.8	45.1	.6	.6	63.4	57
Reunion	22,26,30,34,38	29	-19.2	55.6	1.56	.78	63.9	216
Iran	3,7,11,15,19	34	32.4	54.2	3.82	1.82	62.8	959
Pakistan	2,6,10	38	29.5	69.6	2.3	2.16	63.9	883
Pakistan	12,14	38	30.8	72.1	1.16	.72	63.5	135
Pakistan	18,22	38	27.9	65.2	1.52	1.42	63	311
Pakistan	20,24	38	25.8	68.5	1.32	.62	63.3	126
Pakistan	4,8	38	33.9	74.7	1.34	1.13	64.3	295

Orbit Slots -

Countries	Channels Allocated	Orbit Posn. (degrees)	Boresight Lat. Long. (degrees)		Beamwidth to 3 dB points (degrees)		Centre EIRP (dBW)	Approx. R.F. power (watts)
Maldives	12,16	44	6	73.1	.96	.6	63.7	97
U.S.S.R.	20,24,28,32,36,40	44	44.6	64.3	4.56	2.48	65.4	2840
U.S.S.R.	1,5,9,13	44	58.5	62.4	3.2	1.52	66.3	1503
U.S.S.R.	26,30	44	38.8	59	2.24	1	64	407
U.S.S.R.	12,16	44	38.5	70.8	1.36	.74	64.1	187
U.S.S.R.	18,22	44	41	73.9	1.34	.84	64.5	229
U.S.S.R.	34,38	44	63.3	42	2.64	.84	64.4	442
U.S.S.R.	7	44	61.5	70.1	2.38	.66	67.1	583
U.S.S.R.	3	44	63.5	54.3	1.58	.66	66.9	369
Afghanistan	3,7,11,15	50	35.5	70.2	1.32	1.13	62.8	205
Sri Lanka	2,6,10,14	50	7.7	80.6	1.18	.6	63.6	117
Nepal	17,19,21	50	28.3	83.7	1.72	.6	64.6	215
India	17,19,21,23	56	33.4	75.9	1.52	1.08	64.3	320
India	1,5,9,13	56	11.2	72.7	1.26	.6	63.1	111
India	4,8,12,16	56	25	73	1.82	1.48	63.6	447
India	18,20,22,24	56	16	78.4	2.08	1.38	63.8	498
India	3,7,11,15	56	11.1	77.8	1.36	1.28	63.3	269
India	2,6,10,14	56	19.5	76.2	1.58	1.58	63.5	404
China	2,6,10,14	62	40.5	83.9	2.75	2.05	63.2	853
China	1,5,9,13	62	31.5	88.3	3.38	1.45	62.9	692
China	4,8,12	62	36.3	97.8	2.56	1.58	63.5	655
China	3,7,11	62	27.8	102.3	2.56	1.58	65.1	948
China	22	62	39	104.8	1.48	.6	63.8	154
China	20	62	37.9	101	2.78	.82	63.7	387
China	18	62	35.4	95.7	2.1	1.14	63.4	379
China	16	62	30.2	102.5	1.91	1.23	65.5	603
China	24	62	25.1	101.5	1.86	1.08	65	460
Cambodia	18,20,22,24	68	12.7	105	1.01	.9	64.3	177
India	2,6,10,14	68	25.5	93	1.46	1.13	63.9	293
India	18,20,22,24	68	27.7	79.3	2.14	1.16	63.8	431
India	1,5,9,13	68	22.3	79.5	2.19	1.42	63.3	481
India	17,19,21,23	68	20.5	84.7	1.6	.86	63.6	228
India	3,7,11,15	68	11.1	93.3	1.92	.6	63.4	182
India	4,8,12,16	68	25	86.2	1.56	.9	63.7	238
Bangladesh	15,18,20,22,24	74	23.6	90.3	1.46	.84	63.7	208
Burma	17,19,21,23	74	19.1	97.1	3.58	1.48	63.9	942
Brunei	12,14	74	4.4	114.7	.6	.6	62.5	46
Laos	2,4,6,8,10	74	18.1	103.7	2.16	.78	63.8	292
Mongolia	25,29,33,37,39	74	46.6	102.2	3.6	1.13	64.1	757
Singapore	3,7,11,15	74	1.3	103.8	.6	.6	63.5	58
Thailand	1,5,9,13	74	13.2	100.7	2.82	1.54	63.6	720
U.S.S.R.	26,30,34,38	74	57.6	88.8	3.08	1.68	67.9	2311
U.S.S.R.	32	74	51.7	94	1.52	.6	65.1	213
U.S.S.R.	28	74	63.2	98	1.84	.69	68.1	593
China	15,19,23	80	38	111.8	2.6	1.74	64.9	1012
China	18,20,22	80	27.3	109.4	2.14	1.72	64.5	751

Orbit Slots –

Countries	Channels Allocated	Orbit Posn. (degrees)	Boresight Lat. Long. (degrees)		Beamwidth to 3 dB points (degrees)		Centre EIRP (dBW)	Approx R.F. power (watts)
China	1	80	39.2	116	1.2	.8	64.4	191
China	5	80	37.4	112.2	1.06	.76	64.2	153
China	9	80	41.8	111.4	1.58	1.2	63.6	314
China	21	80	33.9	113.7	1.2	.8	64.3	187
China	24	80	30.8	111.8	1.42	.82	64.7	248
China	12	80	27.4	111.5	1.22	.86	64.4	209
China	14	80	23.8	108.5	1.41	1.08	64.1	283
China	17	80	35.1	108.7	1.42	.88	64.2	238
China	10	80	26.7	106.6	1.14	.94	64	194
Indonesia	2,4,6,8	80	0	101.5	3	1.2	63.3	557
Indonesia	2,4,6,8	80	0	101.5	3	1.2	63.3	557
Indonesia	18,20,22,24	80	-8.1	112.3	3.14	1.46	64.2	873
Indonesia	17,19,21,23	80	-.3	112.3	2.66	2.32	64	1122
Malaysia	16,18,20,22,24	86	4.1	102.1	1.62	.82	63.2	201
Malaysia	2,4,6,8	86	3.9	114.1	2.34	1.12	63.6	434
Vietnam	3,7,11,15	86	16.1	105.3	3.03	1.4	63.4	672
China	3,7,11	92	45.3	122.8	2.5	1.45	65.1	849
China	2,4,6	92	31.1	118.1	2.49	1.69	64.4	839
China	1,5,9	92	21	115.9	2.74	2.42	63.9	1179
China	24	92	41.7	121.1	1.52	.78	64.5	242
China	17	92	43.7	124.3	1.98	.72	64.7	304
China	22	92	48.1	124.8	2.68	.92	65.4	619
China	16	92	36.4	118.5	1.16	.76	64.7	188
China	12	92	33	119.5	1.34	.64	64.4	171
China	10	92	32	117.2	1.2	.74	64.2	169
China	14	92	29.1	120.4	.96	.84	64.3	157
China	8	92	27.4	115.7	1.14	.94	64	194
China	15	92	25.9	118.1	1.02	.84	64.1	159
China	21	92	23.8	121.4	1.14	.82	64.3	182
China	19	92	21.9	112.2	1.84	1.22	63.8	390
China	13	92	12.9	113.7	3.76	2.18	63.6	1360
Australia	1,5,9,13,17,21	98	-18.8	133.5	2.7	1.4	64.3	737
Australia	2,6,10,14,18,22	98	-30.3	135.4	2	1.4	63.2	423
Philippines	16,18,20,22,24	98	11.1	121.3	3.46	1.76	63.7	1034
Indonesia	1,5,9,13	104	-3.2	124.3	3.34	1.94	63.2	980
Indonesia	3,7,11,15,19	104	-3.8	135.2	2.46	2	63.8	854
Japan	1,3,5,7,9,11,13,15	110	31.5	134.5	3.52	3.3	63.2	1758
South Korea	2,4,6,8,10,12	110	36	127.5	1.24	1.02	63.6	209
North Korea	14,16,18,20,22	110	39.1	127	1.3	1.1	64	260
Papua N. Guinea	2,6,10,14	110	-6.3	147.7	2.5	2.18	64.4	1087
U.S.S.R.	19,23,27,31,35,39	110	57.3	112.7	2.67	1.75	64.1	870
U.S.S.R.	25	110	53.4	108.2	2.16	.78	65	385
Caroline Is.	1,5,9,13,17	122	8	149.5	5.36	.77	62.5	531
Guam	2,6,10,14,18	122	13.1	144.5	.6	.6	63.3	55
Marianas Is.	3,7,11,15,19	122	16.9	145.9	1.2	.6	63.5	116
Australia	4,8,12,16,20,24	128	-38.1	145	1.83	1.39	63.3	393

Orbit Slots -

Countries	Channels Allocated	Orbit Posn. (degrees)	Boresight Lat. Long. (degrees)		Beamwidth to 3 dB points (degrees)		Centre EIRP (dBW)	Approx. R.F. power (watts)
Australia	2,6,10,14,18,22	128	-21.5	145.9	2.9	2	63.7	984
Australia	3,7,11,15,19,23	128	-32	147.2	2.1	1.4	64.1	547
New Zealand	13,17,21	128	-41	173	3.3	1.28	64.8	924
Papua N. Guinea	4,8,12	128	-6.7	148	2.8	2.05	63.4	909
Nauru	3,7,11,15	134	-.5	167	.6	.6	62.5	46
New Caledonia	2,6,10,14	140	-21	166	1.14	.72	63.7	139
New Hebrides	3,7,11,15	140	-16.4	168	1.52	.68	62.8	142
U.S.S.R.	20,24,28,32,36,40	140	53.6	138	3.16	2.12	67.7	2857
U.S.S.R.	26,30,34,38	140	55.4	155.3	2.9	2.36	67.9	3057
U.S.S.R.	22	140	65.5	168.5	1.96	.6	68.1	550
Wake Is.	1,5,9,13,17	140	19.2	166.5	.6	.6	63.6	59
Wallis Is.	2,6,10,14	140	-14	-176.8	.74	.6	64.4	88
Marshall Is.	2,6,10,14,18	146	7.9	166.7	1.5	1.5	63.3	348
Fiji	1,5,9	152	-17.9	179.4	1.04	.98	63.7	173
Cook Is.	2,6,10,14	158	-19.8	-161	1.02	.64	64.6	136
Niue	19,23	158	-19	-169.8	.6	.6	64.1	67
New Zealand	1,5,9,13	158	-39.7	172.3	2.88	1.56	63.3	695
Western Samoa	3,7,11,15	158	-13.7	-172.3	.6	.6	63.6	59
Tokelau Is.	20,24	158	-8.9	-171.8	.7	.6	63.8	72
Palmyra	1,5,9,13,17	170	7	-161.4	.6	.6	62.4	45
Samoa	1,5,9,13,17	170	-14.2	-170.1	.6	.6	61.1	33
Tonga	4,8,12,16	170	-18	-174.7	1.41	.68	63.3	148

Index